Contested Waters

This book examines India's transboundary river water disputes with its South Asian riparian neighbours – Pakistan, Bangladesh, Nepal, and Bhutan. It explores the history of disputes and cooperation over the transboundary river water in this region as well as discusses current disputes and future concerns. It analyses how and why existing transboundary river water sharing treaties between India and its South Asian riparian neighbours are confronted with challenges. The book indicates that India's transboundary river water disputes with its South Asian riparian neighbours are likely to escalate in coming years due to the widening of the demand-supply gap in the respective countries. It further shows the impact of bilateral relations on the resolution of transboundary river water disputes, even as cordial relationships do not always guarantee the absence of river water disputes between riparian states. The book looks at some key questions: How political are India's transboundary river water disputes in South Asia? Why do the roots of India's rivers water disputes with Bangladesh and Pakistan lie in the partition of British India in 1947? Why are there reservations against India's hydroelectricity projects or allegations of water theft? Is it possible to resolve transboundary river water disputes among these South Asian countries?

This book will greatly interest scholars and researchers working in the areas of river management, environmental politics, transnationalism, water resources, politics and international relations, security studies, peace and conflict studies, geopolitics, development studies, governance and public administration, and South Asian studies in addition to policymakers and journalists.

Amit Ranjan is Research Fellow at the Institute of South Asian Studies, National University of Singapore, Singapore. He is the author of *India – Bangladesh Border Disputes: History and Post-LBA Dynamics* (2018) and has edited three books – *Partition of India: Postcolonial Legacies* (2019), *India in South Asia: Challenges and Management* (2019), and *Water Issues in Himalayan South Asia: Internal Challenges, Disputes and Transboundary Tensions* (2020). His papers, review essays, and book reviews have been widely published in journals, including *Asian Affairs, Asian Survey, Economic & Political Weekly, India Quarterly, Indian Journal of Public Administration, Studies in Indian Politics, Social Change, South Asia Research, The Roundtable: The Commonwealth Journal of International Affairs, Journal of Asian Security and International Affairs.* He has also contributed commentaries, opinion editorials and reviews in newspapers and websites such as *The Citizen, The Friday Times, The Wire,* and *Prabhat Khabar.*

Amit Ranjan offers an important perspective on water disputes in South Asia triggered by the Partition of the Subcontinent, deepened by the region's inability to depoliticise transboundary river water management and sharpened by the new factors like climate change, and contributes to a better understanding of a major source of regional conflict in South Asia.

C. Raja Mohan is Director at the Institute of South Asian Studies, National University of Singapore.

Contested Waters

India's Transboundary River Water
Disputes in South Asia

Amit Ranjan

Routledge
Taylor & Francis Group

LONDON AND NEW YORK

First published 2021
by Routledge
2 Park Square, Milton Park, Abingdon, Oxon OX14 4RN

and by Routledge
52 Vanderbilt Avenue, New York, NY 10017

Routledge is an imprint of the Taylor & Francis Group, an informa business

© 2021 Amit Ranjan

The right of Amit Ranjan to be identified as author of this work has been asserted by him in accordance with sections 77 and 78 of the Copyright, Designs and Patents Act 1988.

All rights reserved. No part of this book may be reprinted or reproduced or utilised in any form or by any electronic, mechanical, or other means, now known or hereafter invented, including photocopying and recording, or in any information storage or retrieval system, without permission in writing from the publishers.

Disclaimer: The author has made every effort to ensure that the information presented in the book was correct at the time of press, but the author and the publisher do not assume and hereby disclaim any liability with respect to the accuracy, completeness, reliability, suitability, selection and inclusion of the contents of this book and any implied warranties or guarantees. The author and publisher make no representations or warranties of any kind to any person, product or entity for any loss, including, but not limited to special, incidental or consequential damage, or disruption alleged to have been caused, directly or indirectly, by omissions or any other related cause. Perceived copyright omissions if brought to notice will be rectified in future printing.

Trademark notice: Product or corporate names may be trademarks or registered trademarks, and are used only for identification and explanation without intent to infringe.

British Library Cataloguing-in-Publication Data
A catalogue record for this book is available from the British Library

Library of Congress Cataloging-in-Publication Data
A catalog record for this book has been requested

ISBN: 978-1-138-04033-5 (hbk)
ISBN: 978-0-367-54419-5 (pbk)
ISBN: 978-1-003-04512-0 (ebk)

Typeset in Sabon
by Apex CoVantage, LLC

To Papa and Mai

Contents

Figures

Tables

Abbreviations

BC	Boundary Commission
BE	Budget Estimates
BJP	Bhartiya Janata Party
BNP	Bangladesh Nationalist Party
BNU/BTN/NU	Bhutanese Ngultrum or Ngultrum
CAA	Citizenship Amendment Act
CBTE	Cross Border Trade of Electricity
CD	Composite Dialogue
DGPC	Druk Green Power Corporation
GBM	Ganga Brahmputra Meghna
HEP	Hydroelectric Power
HLTC	High Level Technical Committee
ILR	Interlinking of Rivers
IRS	Indus Rivers System
IWT	Indus Waters Treaty
JCE	Joint Committee of Experts
JRC	Joint Rivers Commission
kWh	Kilowatt Hour
MAF	Million Acre Feet
MoU	Memorandum of Understanding
MW	Megawatt
NDA	National Democratic Alliance
NRC	National Register of Citizens
NWDA	National Water Development Agency
SAARC	South Asia Association for Regional Cooperation

Acknowledgements

I am thankful to my MPhil supervisor Professor I.N. Mukherji, who introduced me to the world of contested waters. My PhD supervisor Professor Uma Singh deserves a special thanks for guiding and being supportive to me in various ways.

While working at the Indian Council of World Affairs (ICWA) (2013–January 2017), New Delhi, I started working on this manuscript. Later, in 2017, I moved to the Institute of South Asian Studies (ISAS), National University of Singapore, Singapore (NUS). Here, I managed to satisfactorily finish and submit the draft for publication. I am thankful to all my colleagues at ICWA, particularly Dr Vijay Sakhuja, Dr Pankaj Jha, Dr Samatha, Dr Rahul Mishra, and Dr Shamshad Ahmad Khan.

At ISAS, I am thankful to the Chairman, Ambassador-at-large, Gopinath Pillai, Director, C. Raja Mohan, Dr Dipinder Singh Randhawa, Mr Hernaikh Singh, Dr Rahul, Dr Chulanee, Sylvia, and all other colleagues. I am also thankful to Dr Gyanesh Kudaisya from the South Asian Studies Programme at the NUS and Rajiv Ranjan Chaturvedi from RSIS, Singapore.

I am grateful to my parents who live in a village in Chhapra district in Bihar, India. They do not have an idea what exactly I do, but they firmly believe that I do something worthy. I am also grateful to other members of my joint family.

I am thankful to my wife, Jyoti, who gave me full freedom to do whatever I want to do in my life. My niece, Shibangi, also deserves some amount of thanks for many unknown reasons.

I am also thankful to my friends, Vishwa Negi, Kumar Dhananjay, Neeraj Kumar, Munil, Sandeep Biswas, Somya Nayak, Sudhakar Kumar Rai, Shabbir Raina, Dr Farooq Sulheria, et al.

I acknowledge the help provided by the staff of Jawaharlal Nehru University (JNU) Central Library, Delhi University's Central Reference Library, JNU EXIM Bank library, ICWA library, and NUS Library.

Finally, I am thankful to Shoma Choudhury and the team at Routledge. I also acknowledge Sage for permitting me to publish a part of my paper, "Disputed Waters: India, Pakistan and the Transboundary Rivers" published in *Studies in Indian Politics* (Sage and Centre for the Study of Developing Societies, New Delhi), Volume 4, Issue 2, 2016, pp. 191–205. Besides, some of the facts from my earlier writings on India-Pakistan, India-Bangladesh, India-Nepal, India-Bhutan, and water issues have been inadvertently repeated in this book. But nothing has been fully reproduced.

Introduction

Water from transboundary rivers plays an important role in meeting water-related domestic demands of a large number of countries of the world, though the percentage of such contributions varies from one country to the other. It is estimated that around 39 countries, with a combined population of about 800 million (in 2006), satisfy half of their water demands from sources originating beyond their sovereign territory.[1] Some countries such as Iraq and Syria highly depend on the Tigris and Euphrates Rivers flowing out of Turkey while in South Asia, Bangladesh depends on India for 91% of its water supply.[2]

Globally, there are about 263 major river systems[3] that cross national boundaries and are responsible for 85% of the earth's freshwater runoff.[4] Looking at the spread of river basins one finds that about 145 countries of the world have some part of their territories falling in the international river basins while 21 countries are situated entirely within the international river basins.[5] While majority of the river basins are shared between the two countries, there are many river basins which are shared by more number of countries. For example, 13 river basins of the world are shared between 5 and 8 riparian nations.[6] Five basins, the Congo, Niger, Nile, Rhine, and Zambezi in the African continent are shared between 9 and 11 countries. Then there is River Danube in Europe, which crosses the boundary of 18 countries.[7]

As many countries depend on transboundary rivers to meet their water-related domestic demands, differences, tensions, and disputes over waters from those shared rivers are inevitable. There are two major factors which trigger or may trigger water disputes. First is rising population which proportionally increases the demand for water. According to the *World Water Development Report* of 2018 the world population is expected to be between 9.4 and 10.2 billion by 2050.[8] Consumption-wise, agriculture uses 70% of total waters withdrawal, Industrial activities use 20% and 10% is being used for domestic use.[9] Growing population will need more food, industrial goods, and waters for domestic consumption. On use of water, the *World Water Development Report* further says that the rate of use of water in the world has increased by a factor of six over the past 100 years[10] and continues to grow steadily at a rate of about 1% per year.[11]

To meet the growing demands, there is not enough water. On availability of water, the *World Water Development Report* projects that by 2050 many regions of the world may turn into water-scarce areas or those which are already facing water scarcity may face severe water scarcity zone. In this report a region is defined as water scarce when total annual withdrawals for human use is between 20 and 40% of the total available renewable surface water resources, and it can be defined as a severely water scarce area when such water withdrawals exceed 40%.[12] To measure water scarcity, there is Falkenmark Water Index, developed by Malin Falkenmark.[13] This index defines and measures water scarcity in terms of per capita of water availability. According to Falkenmark Index the availability of 1,700 cubic metres of water per person in a country makes it water stressed. Below it, availability of 1,000 cubic metres of water per person represents a state of "water scarcity" – and below 500 cubic metres means that the region or country is facing "absolute scarcity".[14]

The second factor which affects water availability is the phenomenon of climate change. This, often, creates a situation where a part of the region in a country is reeling under drought or facing drought-like conditions while, simultaneously, there are floods in other parts. If the climate emergency situation into which the world has entered now is not being effectively addressed, there are going to be more natural disasters in coming days. In some parts of the world, the climate emergency situation is getting worse. To address the situation, at the United Nations (UN) Climate Action Summit 2019 held at New York, UN Secretary-General António Guterres called on all leaders to take bold actions to make carbon-neutral world by 2050.[15] Despite such climate emergency situation which the world is facing, for many countries it still remains business as usual. So far, only Scotland has taken strong steps to curtail the impact of climate emergency. The country has declared a climate emergency with an aim to reduce greenhouse gas (Carbon dioxide, etc.) emissions to net-zero by 2045.[16]

The International Panel on Climate Change, an intergovernmental body set up by the United Nations in 1988 to study the impact of climate change, in its 2018 report, has highlighted the fact that an increase in global warming from 1.5 degree Celsius would make some of the coastal cities of the world, more vulnerable to sea floods. Simultaneously, there will be drought or drought-like situation in some of the areas due to low precipitation.[17] Climate change will also change the weather pattern in many regions due to which one may experience rain during what is generally regarded as non-rainy seasons, and extended number of winters or hot days.

In both situations, floods and drought-transboundary waters may trigger tensions and disputes, if the countries do not cooperate to jointly address the situation. During floods the upper riparian may release or be accused of releasing more than the required waters into another country to secure its area or during water shortage situation the same riparian may hold or divert the waters in its region instead of releasing for use by the lower riparian States. In fact, such situations prevail even now. China, upper riparian to India's Brahmaputra River, is being accused by India for floods during monsoon period in the river flowing in India's northeastern States. Dam breaches in the upper riparian, as reported, causes severe floods in the lower riparian areas.[18] Pakistan, a lower riparian to India, accuses India for not releasing or diverting waters from the transboundary rivers, mainly, during the summers.[19]

To address the issue of water sharing between the riparian States there are globally accepted principles, rules, and laws whose efficacy to de-escalate the tensions and manage disputes would be highly tested in the coming years when many parts of the world are expected to enter into scarcity or severe water scarcity situation. Before reaching any such conclusion, it is imperative to know about the rules, principles, and laws on transboundary waters.

Water laws, doctrines, and principles

Disputes over water from transboundary rivers has a history whose roots lay in the ancient time, as mentioned in the following section. In the ancient world there were codified laws to tackle the then prevailing situation. Many of the present international laws have evolved from the laws in ancient times. King Hammurabi (1795–1750 BC) of Babylonia codified all aspects of Babylonian laws in a work known as the Code of Hammurabi. This code also included laws on water management, irrigation, etc. In ancient Rome in AD 528 Roman king Justinian I (AD 483–565) ordered a compilation of all existing laws which were evolved in the 13 centuries. This is known as the Justinian Code (*Corpous Juris Civilias* or Body of Civil Law).[20] The Justinian Code also included laws on waters. The riparian (from the Latin word *ripa* meaning bank or shore) doctrine, also called common law, on water developed in the Justinian code and provide framework for water allocation throughout the Roman empire. The Visigoths, German invaders in Spain, after their success established rules by royal order that prohibited construction of weirs or dams that could hinder movements of fish or navigation. By the end of the ninth century water allocation policies and rules evolved into an extensive law.[21] In the medieval period, an important development was the French Revolution of 1789. After the revolution, in 1804 the Code Napoleon (*Code Civil des Francais* or French Civil Code) was compiled under Napoleon I (1769–1821). This code defined riparian water rights, navigation rights, etc. The Napoleon Code provided guidelines on which water laws and riparian rights were later developed in the United States of America.[22]

Over time, such laws on water and riparian rights were customised[23] and, in modern times, the international organisations have codified water laws and principles so that disputes between riparian States could be managed and rights of the lower riparian States remain secured. There are also water doctrines at the country level which are often used to manage their internal water disputes. Due to their effectiveness in managing and addressing water disputes, some of those doctrines have acquired a global reputation and are used in other countries too.

In most countries water rights are either tied to the land under control or to the time of initial use. In the United States of America these are called "riparian" and "prior appropriation" doctrines, respectively.[24] Under the riparian doctrine, the owner of land contiguous to a stream has use rights over its water. In its basic form, consumptive use is excluded, and the owner of land has to return comparable quality and quantity of waters to the source. However, over time, "reasonable use" modifications have been allowed for consumptive uses such as irrigation.[25] Dispute under this principle often involves intervention from a third party, such as government. Second, there is a principle of prior appropriation. It accords the users water rights in the order in which the water was originally withdrawn from the source.[26] Third, there is a principle of "the community of interest theory" according to which the entire basin is taken as a single economic unit irrespective of boundaries, and rights over the waters are vested in the community of co-riparian States which has to be utilised to the maximum benefit of all in an integrated manner.[27] Fourth, a "public interest" doctrine has evolved in the United States. Under it, water diversions to other place are being blocked if it causes significant harm to residents or ecosystems.[28]

In case of transboundary river waters or border waters disputes there are mainly four doctrines:[29]

Absolute Sovereignty: It combines the features of riparian and prior appropriation doctrines by claiming a right to consumptive use based on location. This doctrine took the most notorious form in the Harmon Doctrine which has been named after an 1895 opinion given by Attorney General Harmon of the USA on the River Rio Grande water dispute between the USA and Mexico. In this opinion, the United States denied that it had any obligation to guarantee stream flow to downstream Mexico, even from return discharges. This doctrine has little legitimacy in international law.

Territorial Integrity Theory or Natural Water Flow Theory: It entitles every lower riparian to the natural flow of the river without any interference from the upper riparian. Any form of interference is considered a violation of the territorial integrity of the lower riparian of which the river is a constituent. It was advanced by Egypt in regard to the dispute with Sudan on sharing of water from the River Nile water.

Historical Use: It basically means "first in time, first in right". It accords the first user of water a priority right whether or not his land is contiguous to the stream, or downstream from other users. At the level of countries this doctrine corresponds to prior appropriation rights.

Equitable Distribution or Equitable Utilisation: It regards an entire basin area as an indivisible unit which has to be developed for the benefit of the maximum number of people regardless of the territorial boundaries.

Taking into account the mentioned doctrines, some rules have come into effect. On the basis of previously mentioned principles, laws have evolved in the modern times to share river waters between the riparian States. One of the earliest was Helsinki rule of 1966 which has incorporated the notion of "prior appropriation".[30] Years later, the Helsinki convention on Transboundary Waters was adopted by the Senior Advisors to the European Economic Commission for European governments.[31] It was opened for signature from 17 to 18 March 1992 at Helsinki and was, further, opened for signature at the United Nations Headquarters in New York until 18 September 1992. It came into force in 1996, and amendments were made in 2003 in articles 25 and 26. The amended articles entered into force on 6 February 2013.[32] Some of the provisions regarding transboundary rivers in the Helsinki convention (mostly written in verbatim) are as follows:[33]

1   The basin States shall take all appropriate measures: (a) To prevent, control and reduce pollution of waters causing or likely to cause transboundary impact; (b) To ensure that transboundary waters are used with the aim of ecologically sound and rational water management, conservation of water resources and environmental protection; (c) To ensure that transboundary waters are used in a reasonable and equitable way, taking into particular account their transboundary character, in the case of activities which cause or are likely to cause transboundary impact; (d) To ensure conservation and, where necessary, restoration of ecosystems.

2   To prevent, control and reduce transboundary impact, the basin States shall develop, adopt, implement and render compatible relevant legal, administrative, economic, financial and technical measures, in order to ensure that: (a) The emission of pollutants is prevented, controlled and reduced at source through the application of, inter alia, low- and non-waste technology; (b) Transboundary waters are protected against pollution from point sources through the prior licensing of wastewater discharges by the competent national authorities, and that the authorized discharges are monitored and controlled; (c) Limits for waste-water discharges stated in permits are based on the best available technology for discharges of hazardous substances; (d) Stricter requirements, even leading to prohibition in individual cases, are to be imposed when the quality of the receiving water or the ecosystem so requires; (e) At least biological treatment or equivalent processes are to be applied on municipal waste water; (f) Appropriate measures are taken, such as the application of the best available technology, in

order to reduce nutrient inputs from industrial and municipal sources; (g) Appropriate measures and best environmental practices are to be developed and implemented for the reduction of inputs of nutrients and hazardous substances from diffuse sources, especially where the main sources are from agriculture; (h) Environmental impact assessment and other means of assessment are to be applied; (i) Sustainable water-resources management, including the application of the ecosystems approach, is to be promoted; (j) Contingency planning is to be developed; (k) Additional specific measures are to be taken to prevent the pollution of groundwaters.

3　The agreements or arrangements mentioned in the Helsinki convention provide for the establishment of joint bodies whose task shall be: (a) To collect, compile and evaluate data in order to identify pollution sources likely to cause transboundary impact; (b) To elaborate joint monitoring programmes concerning water quality and quantity; (c) To draw up inventories and exchange information on the pollution sources; (d) To elaborate emission limits for waste water and evaluate the effectiveness of control programmes; (e) To elaborate joint water-quality objectives and to propose relevant measures for maintaining and, where necessary, improving the existing water quality; (f) To develop concerted action programmes for the reduction of pollution loads from both point sources (e.g. municipal and industrial sources) and diffuse sources (particularly from agriculture); (g) To establish warning and alarm procedures; (h) To serve as a forum for the exchange of information on existing and planned uses of water and related installations that are likely to cause transboundary impact; (i) To promote cooperation and exchange of information on the best available technology as well as to encourage cooperation in scientific research programmes; (j) To participate in the implementation of environmental impact assessments relating to transboundary waters, in accordance with appropriate international regulations.

4　The Riparian States shall establish and implement joint programmes for monitoring the conditions of transboundary waters, including floods and ice drifts, as well as transboundary impact. They shall agree upon pollution parameters and pollutants whose discharges and concentration in transboundary waters shall be regularly monitored. At regular intervals, there is a need to carry out joint or coordinated assessments of the conditions of transboundary waters and the effectiveness of measures taken for the prevention, control and reduction of transboundary impact.

Another important rule was adopted years later at the 71st meeting of the International Law Association at Berlin in August 2004. It is a comprehensive rule on water.[34] In 1996 when the Water Resources Committee (WRC) presented its first report at the 1996 Helsinki Conference of the Association, at that time, the Association, on the recommendation of WRC, adopted two sets of articles: The Articles on Private Law Remedies for Transboundary Damage in International Watercourses and the Supplemental Rules on Pollution.[35] The Association also approved the resolution proposed by the WRC on the further consideration by the General Assembly of the United Nations of the Draft Articles on the Law of the Non-Navigational Uses of International Watercourses proposed by the International Law Commission and requested the Secretary General of the Association to transmit the Resolution to the Secretary General of the United Nations Organization with the suggestion that it is circulated to members of the Organization.[36] Simultaneously, the WRC continued with its work and prepared its Second Report, for the Millennial Conference in London in 2000. At the Millennial Conference, the WRC presented the Campione Consolidated Rules and an article on Adequate Stream Flows, and a progress report on a project to revise and update the Helsinki Rules and the other rules supplementing the Helsinki Rules.[37] The Campione Consolidation was a compilation and consolidation of the prior work of the WRC as approved by the Association over a period of 34 years.[38] After a few rounds of meetings, the Berlin rules were finally adopted by the association. Some of the major provisions (written in verbatim) of the Berlin rules on transboundary freshwater resources are as follows:[39]

1 Basin States shall in their respective territories manage the waters of an international drainage basin in an equitable and reasonable manner having due regard for the obligation not to cause significant harm to other basin States.
2 Basin shall develop and use the waters of the basin in order to attain the optimal and sustainable use thereof and benefits therefrom, taking into account the interests of other basin States, consistent with adequate protection of the waters.
3 In determining an equitable and reasonable use[40], States shall first allocate waters to satisfy vital human use.
4 Allocation by agreement or otherwise to one basin State does not prevent use by another basin State to the extent that the basin State to which the water is allocated does not in fact [make] use of the water.
5 Use of a water includes water necessary to assure ecological flows or otherwise to maintain ecological integrity or to minimize environmental harm.
6 Use of water by a basin State other than the one to which the water is allocated does not preclude the basin State to which the water is allocated from using the water when it chooses to do so.
7 Basin States, in managing the waters of an international drainage basin, shall refrain from and prevent acts or omissions within their territory that cause significant harm to another basin State having due regard for the right of each basin State to make equitable and reasonable use of the waters.

As an international organisation, the United Nations (UN) has also looked at the issue of transboundary waters issues which have been a reason for disputes and conflicts between a number of riparian States. In 1997 the UN drafted "Non-Navigational uses of International Watercourses". It ensures the utilisation, development, conservation, management, and protection of international watercourses. It talks about promoting their optimal and sustainable utilisation for present and future generations. The 1997 UN law has various provisions calling the transboundary river States to cooperate by exchanging information about data flow, etc. and taking appropriate measures to prevent the causing of significant harm to other watercourse states.[41]

In its 63rd session, the UN General Assembly adopted a resolution on the Law of Transboundary Aquifers by consensus on 11 December 2011. This resolution encourages States "to make appropriate bilateral or regional arrangements for the proper management of their transboundary aquifers. The provisions under it include cooperation among States to prevent, reduce, and control pollution of shared aquifers".[42] Year 2013 was celebrated as an International Year of Water Cooperation. To attract attention of people and ways to address the challenge, the UN publishes monographs, reports such as *World Water Development Report*, etc.

Transboundary river water disputes

Despite all such doctrines, principles, and rules, transboundary and border water disputes remain a major issue between the riparian States. Most of the upper riparian States emphasise the principle of *Absolute Territorial Sovereignty*. They are guided by the water nationalism, and there is an absence of political will as well as a lack of trust between and among the riparian States.[43] Besides, most of the powerful riparian States try to avoid or are not a party to the important international conventions on transboundary river waters. For example, India and China are not yet parties to the UN Water Convention (UNWC) of 1997. China's reason for not signing the UNWC are:[44] the 1997 UNWC overemphasised downstream States rights at the expense of the interests of upstream States; Part III of the 1997 UNWC on Planned Measures triggered national security and territorial sovereignty concerns from China; and China has an objection to the mandatory settlement of disputes through mediation or conciliation process by a third party. Unfortunately, at the time of signing of the UNWC, China had a setback on South China Sea by the Arbitration Court which, further, resolved its stance against entering into such arrangement.[45] Reasons for India's objection are:[46]

1   Article 3 of the UNWC does not adequately reflect "the principle of freedom, autonomy, and the rights of States to conclude international agreements on the international courses without being fettered by the present Convention".[47]

2   Article 5, dealing with equitable and reasonable utilization and partici-
    pation, is ambiguously drafted, vague and so difficult to implement.
3   Article 32, dealing with non-discrimination, calls for political and eco-
    nomic integration among States of the region. However, as all water-
    course regions are not so integrated, this provision will be difficult to
    implement in certain regions.
4   Article 33 deals with peaceful settlement of disputes. India's position is
    to let the parties engaged in such disputes choose a procedure of engage-
    ment with mutual consent.

Looking at the transboundary water disputes, Ismail Serageldin, Vice President of the World Bank in the early years of the twenty-first century stated
that the wars in this century could be over shared waters unless they are
appropriately managed and ways to manage them change.[48] Since then, a
number of world leaders and head of the international institutions have
repeated such predictions. Although disputes have surfaced, war over the
waters is yet to become a reality.

Historically, in ancient times wars over waters have taken place and, also,
water has been used as an instrument in the wars. In around 2600 BC the
city states of Umma and Lagash (including Girsu, the religious capital) in
Mesopotamia (modern Iraq) fought for centuries to establish their hold over
irrigation canals fed by the River Tigris.[49] There are also examples of water
being used as an instrument during the wars. In 1573 AD, at the beginning of
the 80-year war against Spain, the Dutch flooded the land to break the siege
of the Spanish troops on the town Alkmaar. The same defence was used to
protect Leiden in 1574. This strategy is known as the Dutch Water Line and
was used frequently for defence in later years.[50] In 1937–38 Chang Kai-Shek
ordered his army to destroy flood control dikes along the Huang He (Yellow)
River to flood areas threatened by the Japanese army.[51] In modern times,
Israel has occupied most of the rich water sources in the West Asia and it has
deliberately failed in peace talks with Syria over the issue of Lake Tiberia.[52]
Then, although India has warned Pakistan of stopping waters flowing into it
through the upper riparian, after a militant attack in 2016 in the Uri sector
of the Indian side of Jammu & Kashmir in which 20 Indian Army personals
lost their lives, it has taken any such step except speed up the works on
ongoing projects to use waters from the eastern rivers (Sutlej, Ravi and Beas)
allocated to it under the Indus Waters Treaty (IWT).[53]

According to the Human Development Report of 2006, published by the
United Nations Development Programme, transboundary water disputes
prevails because of:[54]

(a) *Competing claims and perceived national sovereignty imperatives:*
    Many countries remain deeply divided in the way they look at the
    shared waters. For example, India sees the flow of the Brahmaputra
    and Ganges Rivers as a national resource while Bangladesh looks at

the same water as a resource that it has claims to on the grounds of prior use patterns and needs. Such differences are directly related to the claims that both countries see as legitimate and necessary to their national development strategies. Competing claims over shared waters and national sovereignty imperatives also construct water nationalism in the respective riparian States which, often, makes any effort for cooperation difficult.

(b) *Weak political leadership:* Political leaders are accountable to their respective domestic constituencies and not to the basin-sharing communities from other sovereign state(s). They assure the domestic constituency to secure their water interests which may cause harm to the communities living in the same catchment areas across the border. They do so because water sharing treaty does not have any political incentive. In such regions, an idea of more equitable water sharing might be good for human development in a basin, but it might be a vote loser at home. Hence, the leaders shy away from making such proposals. Growing "water nationalism" has forced even the powerful leaders to not take any chance on drafting and implementing equitable and reasonable water sharing treaty, though few like Sheikh Hasina, Prime Minister of Bangladesh has, largely, managed to keep public opinion on water sharing with India in her favour.

(c) *Asymmetries of power:* In many basins, rivers flow through countries marked by large disparities in wealth, power, and negotiating capacity. These disparities shape the willingness among the powerful riparian States to cooperate, negotiate, and share benefits of the shared waters. However, the weaker riparian States are not very enthusiastic because they have to, largely, depend on the powerful countries from the region for favourable cooperative arrangements.

(d) *Nonparticipation in basin initiatives:* Perceptions of the benefits of participating in multilateral basin wide initiatives influences a country's decision to either become a part of it or not. After calculating their cost-benefit analysis, some of the powerful riparian States of the region have decided to remain absence from basin wide initiatives. Absence of such relatively power riparian States, in basin wide initiatives makes such group weak and less effective. For example, China's absence from the Mekong River Commission is seen by some parties as a source of potential weakness of the commission.

Over the years, disputes over transboundary waters have become difficult to resolve because states in international system, as neo-realists argue, are rational actors who, mainly, shy away from cooperation. They remain fearful that that if they were to cooperate, their partners could eventually turn out to be better off than them by virtue of having achieved relatively greater gains.[55]. Also, Kenneth Waltz and Scott Sagan maintain that rational States do not likely run into major risks for minor gains.[56] States, according to neo-realists, mainly, want to defend their position in the international system; therefore, it, often, becomes difficult to achieve cooperation. Joseph Grieco characterises such States as "defensive positionalists".[57] However, if there is substantial pay-off and such cooperation does not affect their position and that of others, mainly of the rival in the system, States, as neoliberals find, even in anarchical systems, do cooperate.

Robert Keohane finds that cooperation is possible even between the rival States. He writes

> cooperation requires that the actions of separate individuals or organi-zations – which are not in pre-existent harmony – be brought into con-formity with one another through a process of negotiation, which is often referred to as "policy coordination". Cooperation occurs when actors adjust their behavior to the actual or anticipated preferences of others, through a process of policy co-ordination.[58]

For Keohane, cooperation does not imply an absence of conflict. On the contrary, it is typically mixed with conflict and reflects partially success-ful efforts to overcome conflict, real or potential. Cooperation takes place only in situations in which actors perceive that their policies are actually or potentially in conflict not where they are in harmony. Cooperation should not be viewed as the absence of conflict but rather as a reaction to conflict or potential conflict. Without spectre of conflict there is no need to cooperate.[59]

Earlier, the functionalist, David Mitterney wrote,

> Cooperation is possible when there is peace, which can be attained only if the intrusion of power politics was checked, have sovereignty sacri-ficed and efforts made toward material unity in an increasingly interde-pendent world.[60]

Taking into account the larger benefits of cooperation, region and theo-ries on regionalism evolved. Many countries, especially the smaller ones, have drawn benefits by becoming a part of one or the other regions and their political and economic systems. However, with the growing nationalism and protectionism, the idea of region and regionalism are facing severe setback. In recent times the idea of Regionalism has faced its first major challenge when, after a referendum, the United Kingdom decided to pull out from the European Union in 2016.

Constructivists and Marxists theories have different understanding on cooperation. Alexander Wendt debates against the prevailing anarchical system about which the neo-realists talk. Wendt finds that anarchy is basi-cally a construct of nation-states.[61] For Wendt, the structures of human asso-ciation are determined primarily by shared ideas which construct identities and interests of the actors. The Constructivists' idea of cooperation under-pins shared ideas and interests. Anarchy, as defined by the neo-realists, has been also criticised by the Marxist scholars whose focus is on the relation between the base and superstructure. For Marxist scholars it is a capitalist mode of production which defines relations between the States. Gramscian and neo-Gramscian schools look at society-state complexity, class structure, and its relations with similar class structures from other countries.[62]

Among all the mentioned theories, continuous wars, tensions, and the continuous changes in global affairs have popularized neo-realism more than any other. To defend their position in international system and maxim-ise security, States plan strategies even at the cost of jeopardizing the others. Quite often, they do so even without giving required calculations on spillo-ver effects on them. For example, in the 1980s United States of America trained a number of Afghans and Pakistani civilians to fight against its Cold War (1948-1991) rival Soviet Union which invaded Afghanistan in 1979 to secure its position and interests in the Central Asian country. A few years after the invasion, in 1991 the Soviet Union was disintegrated and in 2001 the USA was attacked by militants from Afghanistan. Post-2001, the USA launched an attack on Afghanistan and is yet to completely resolve the issue of militancy there.[63] Due to such invasions, fight against the invaders, and attacks, Afghanistan has turned into one of the most disturbed and insecure countries in the world.

In terms of shared waters, to achieve water security, most of the States want to keep maximum quantity of transboundary waters under their control. They, largely, believe in the neo-realist theoretical assumptions that any water sharing cooperation may disturb their water interests and positions vis-à-vis their co-riparian neighbour(s). They hardly care for the water-related interests of the other riparian States unless it damages their relationships with that State. On contrary, in some cases, the States are ready to even accept "uncomfortable" demands of others or make a com-promise with other riparian States for higher gains. For example, the Union government of India is pushing for implementation of the water deal on River Teesta with Bangladesh despite opposition from the Indian state of West Bengal. India wants to have a very close relationship with Bangladesh for economic and security reasons.

Fredrick Frey, Thomas Naff, and Ruth Maston developed water conflict prediction theories in 1980s. They have used three different approaches to predict such conflicts which succeed disputes: the typology of conflict, perceptions of conflict, and cognitive mapping. The framework they pro-pose is called *interest-position-power matrix*. The main hypothesis the three authors developed is that the potential for a water conflict is highest when the ranking of States according to the interests-position-power matrix is near equal, which implies that conflict is more likely to occur when interests and power of the riparian countries are at equal level.[64] In South Asia, this matrix aptly applies and explains India-Pakistan disputes over their shared rivers. Both are nuclear powers and heading towards facing water scarcity.

In some cases, water is also looked at as a source or means of power of one riparian State over the others. By having control over the waters, the upper riparian can regulate the lower riparian and inflict damages on it. For example, in 1982 Israel launched a massive operation against Lebanon to seize control of the River Litani. This was then regarded important for Israel's national security. In the water stressed West Asia, unimpeded access to water is a great advantage.[65] By having Litani under it, Israel could enjoy more power over Jordon. In the case of Israel-Palestine, in addition to partially controlling the water resources in West Bank, Israel also controls the water abstractions across the entire Western Aquifer Basin.[66]

However, despite tensions and disputes, countries do cooperate or try to do so after realising the benefits from such cooperation. In the last 60 plus years there have been only 37 acute disputes involving violence, compared to 150 treaties signed between the riparian States. Globally, since 805 AD, according to an estimate by Food and Agriculture Organisation, about 3600 treaties on international water course have been drawn up.[67] Even some of the world's most vociferous enemies have negotiated successful water agreements and, in some cases, the institutions they have created often proved to be resilient, even when bilateral relations have been strained or tensed. For example, the Mekong committee – established by the governments of Cambodia, Laos, Thailand. and Vietnam in 1957 – shared data and information on water resources development, even during the Vietnam War (1955-1975). Israel and Jordan have held secret "picnic table" talks to manage the Jordan River. As a result, in 1994, they signed a treaty to share the water from Jordan River.[68] IWT 1960 too has survived major wars (1965, 1971 and 1999) between India and Pakistan. To look at the disputes/conflicts and cooperation over transboundary river waters, some of the students from Oregon State University took up a research project study in which they concluded that violence over shared waters is strategically irrational, hydrologically ineffective, and not economically viable.[69]

India's water disputes with its South Asian riparian neighbours

In India, a broader definition of river was adopted at the first India Rivers Week held on 24–27 November 2014 in New Delhi, where it was accepted:

> a river is more than a channel carrying water; it is also a transporter of sediment; it is also the catchment, the river bed, the banks, the vegetation on both sides, and the floodplain. The totality of these constitutes a river. A river harbours and interacts with innumerable organisms (plant, animal and microbes). It is a natural, living, organic whole, hydrological

and ecological system, and part of a larger ecological system. A river is also a network of tributaries and distributaries spread over its basin and the estuary.[70]

Hence, river is not only about waters but beyond that.

In South Asia, India shares its waters with four of its neighbours – Pakistan, Bangladesh, Nepal, and Bhutan. India's water relations with the South Asian riparian neighbours are different. With Pakistan and Bangladesh, the disputes are mainly over the sharing of waters from the transboundary rivers. In the case of Nepal, tensions simmer because of floods in the Indian state of Bihar from the rivers flowing into the state from Nepal and Hydro-electric Power (HEP) projects. India and Bhutan have issues, mainly, over HEP projects in the Bhutanese territory.

More than the water relationships, in South Asia or in other regions of the world, political relations the countries share define the degree of transboundary water disputes and the nature of cooperation over it. In the case of India and Pakistan, the history of their rivalry which has led to four wars (1947–48, 1965, 1971, and 1999) between them influences their transboundary water-related policies and behaviours. Due to their political rivalries the two countries, despite having provisions of cooperation under the IWT of 1960 between them, most of the time use the clauses of treaty to create hurdles in each other's water-related projects. Even though the IWT is intact, many people in both India and Pakistan believe that it does not look at their country's interests. Unfortunately, such numbers are growing and many from both countries, especially from India, call to review or better abrogate the IWT. In 2016, after the militant attacks in Uri sector of the Indian side of Jammu & Kashmir and again in February 2019, the Union Government of India talked about stopping the flow of waters to Pakistan. In 2016 the Indian Prime Minister Narendra Modi made a statement that water and blood cannot flow simultaneously[71] and in 2019 his cabinet colleague and then Union Minister of Water Resources, Nitin Gadkari, said that his government is coming out with a plan and infrastructures to stop water flows from Eastern flowing rivers, allocated to India under IWT, into Pakistan.[72]

Unlike India-Pakistan, India and Bangladesh share good political relationships. The two countries share 54 rivers between them and first entered into a water-related dispute when the Farakka barrage was proposed in independent India. At that time Bangladesh was East Pakistan and the Government of Pakistan's contention was that the Farakka barrage would affect the water flows from River Ganges into then East Pakistan. After East Pakistan was liberated in 1971, gradually, India and Bangladesh resolved their bilateral differences over the Farakka barrage. In 1977 the two countries agreed on sharing waters from the River Ganges and in 1996 India and Bangladesh signed the Ganges Water sharing treaty. However, they could not achieve any breakthrough on Teesta waters, which the two governments agreed to in 2011. The Union government of India maintains that the onus lies on the government of West Bengal to implement the agreement. The Mamata Banerjee headed West Bengal government feels that if the agreed amount of water is released to Bangladesh, it would harm the water interests of the farmers from her State. On Centre-State waters, particularly on transboundary rivers, final constitutional authority lies with the Central government. Article 253 of the Indian constitution says that

> Parliament has power to make any law for the whole or any part of the territory of India for implementing any treaty, agreement or convention with any other country or countries or any decision made at any international conference, association or other body.[73]

However, the Union government has cleared its position in 2015 that it would not move ahead on Teesta waters without taking West Bengal's government on board.

Like many other parts of the world, it is suggested that one of the ways to mitigate the disputes and cooperate on the transboundary rivers waters is to consider them a common or shared resource instead of a sovereign property owned by one or the other riparian. Also, efforts toward having a basin level management of the transboundary rivers have never taken place in the region, although the now defunct regional organisation – South Asia Association of Regional Cooperation – at its 15th summit held at Colombo, Sri Lanka in 2008 adopted a declaration stating that

> The Heads of State or Government expressing their deep concern at the looming global water crisis, recognized that South Asia must be at the forefront of bringing a new focus to the conservation of water resources. For this purpose, they directed initiation of processes of capacity building and the encouragement of research, combining conservation practices such as rain water harvesting and river basin management, in order to ensure sustainability of water resources in South Asia.[74]

Unfortunately, no concrete step has been taken in such a direction by the member countries.

Transboundary waters and HEP projects are two different issues, however, like many other close riparian regions of the World, in South Asia also they are interlinked. The HEP projects affect the flow of the transboundary rivers because waters are collected to keep the water plants running throughout the year. In South Asia, Nepal and Bhutan are water rich countries but do not have adequate technology to generate hydroelectricity. They are being aided by India. Over the years, India has developed several HEP projects in Bhutan and Nepal. Instead of transboundary waters, India has issues with Nepal and Bhutan over the HEP projects. There is mixed feeling on India's assistance to Nepal and Bhutan in the development of HEP projects. Initially, with little suspicion, India's investment in this sector was welcomed. However, over time, there has been a growing feeling, among a sizeable number of people from these riparian States, that India "exploits" their water resources for its own benefits. There has also been an increasing number of concerns over the India-aided HEP projects. For example, the guidelines issued by the Indian Cross Border Trade of Electricity (CBTE) in 2016 put the Nepali and Bhutanese governments in a confused and difficult situation. Clause 5.2.1 of the guidelines stated,[75]

> Considering that electricity trade shall be involving issues of strategic, national and economic importance, participating entities (Participating Entity(ies)) complying with following conditions shall be eligible to participate in cross border trade of electricity after obtaining one-time approval from the Designated Authority:

a   Import of electricity by Indian entities from Generation projects located outside India and owned or funded by [the] Government of India or by Indian Public Sector Units or by private companies with 51% or more Indian entity (entities) ownership;
b   Import of electricity by Indian entities from projects having 100% equity by Indian entity and/or the Government/Government-owned or controlled company (ies) of [a] neighbouring country.
c   Import of electricity by Indian entities from licenced traders of neighbouring countries having more than 51% Indian entity(ies) ownership, from the sources as indicated in para 5.2.1 (a) and 5.2.1(b) above.
d   Export of electricity by distribution licensees/Public Sector Undertakings (PSUs), if surplus capacity is available and certified by the concerned distribution licensee or the PSU, as the case may be.

After Nepal and Bhutan expressed deep dissatisfaction over such provisions, in 2018 amendments were made in that CBTE's document.[76] The 51% rule has been removed now. Thereafter, in Bhutan works started in few of the projects, such as in Kholongchhu hydroelectric project. Under amended provisions Import and Export can be carried out after approval from the Designated Authority. Under Clause 6.1 of the amended act:[77]

> The Designated Authority shall grant approval for export/import of electricity only after taking into account the generation capacity (as available) and the demand. Imports may normally be permitted only when the demand exceeds generation capacity (as available) in the country; and Exports may normally be permitted in case of capacity being in excess of the domestic demand. However, Govt. of India reserves

the right to import/export electricity from/to neighbouring countries for reason of larger policy interests.

Provision 6.2 of the Amended Act says that:[78]

The Designated Authority shall consider the application for approval of participating Entity(ies) only after the receipt of the equity pattern of ownership of the said Entity(ies) along with other details as prescribed by the Designated Authority. In case where there is a change in the equity pattern, the participating Entity shall intimate the Designated Authority within thirty days from such change in equity pattern for continuation of the approval.

Why this book?

The major reason for planning this book is to understand, examine, and analyse India's water disputes with its South Asia's co-riparian States. It looks at the history and politics of the disputes and cooperation. While doing so an attempt has been made to fill the gap left by previous literature on this issue which have been mainly in the form of research papers, monographs, and articles.

The book *Water: Asia's New Battleground* envisages water and river beds as a new battleground in Asia. In this, the author, Brahma Chellaney, argues that the battles of tomorrow may be over water.[79] *International Conflict Over Water Resources in Himalayan Asia* focuses upon the transboundary river systems and basins, and on the freshwater crisis in Himalayan Asia. It also looks at the growing competition among the Himalayan Asian countries to establish their control over the transboundary waters, which are reasons for the growing disputes.[80] *Riverine Neighbourhood: Hydropolitics in South Asia* was published in 2016. This report by Manohar Parrikar Institute of Defence Studies task force looked at the hydropolitics, cooperation, and competing claims over transboundary waters among the South Asian States. It also finds that the water remains "political" in a South Asian context.[81] *Water Issues in Himalayan South Asia: Internal Challenges, Disputes and Transboundary Tensions* is an edited book. This book looks at the water issues in the respective South Asian countries depending upon Himalayan River system and China, and then talks about how the domestic water pressure is creating tensions over the transboundary waters.[82] *India's Water Security Challenges: Myths, Reality and Measures* discusses the water-related complexities in South Asia. It also looks at the increasing demand for water in South Asia because of the rising population, rapid industrialisation, and inefficient water management system.[83] In addition to them there are chapters in edited books or articles in the journal or short essays written on this theme. *Three River Waters Treaty* is a study of India's waters treaty with Pakistan, Bangladesh, and Nepal. In this chapter in an edited book, author Ramaswamy R. Iyer has discussed the merits and shortcomings of those treaties.[84] This author has also published a journal article on the theme. *In Water Conflicts in South Asia: India's Transboundary Water Conflicts with Bangladesh, Pakistan and Nepal*. I have investigated the history and politics of existing water disputes India has with its three riparian neighbours.[85] Besides the mentioned works there are many other writings on water issues in the individual country and their intricate relations with the transboundary water disputes.

Different from the earlier works, this book looks at the history and politics of the India's transboundary river water disputes and at the HEP projects related differences and controversies India has with Nepal and Bhutan, in particular. In this book I have made three arguments and have attempted to prove them. First, riparian States sharing good or not good political relationships have signed treaties, agreements or Memorandum of Understandings to address their water issues; however, disputes between them remain or surface after a few years of signing such arrangements. In many cases where the riparian States do not share good political relationships, treaties have been signed after mediation by a third country or by an international organisation. An example of the latter case is that the IWT between India and Pakistan in 1960 was mediated by the World Bank. There are also examples of existing water disputes or differences even between the countries who have close political relationships, such as on River Teesta waters between India and Bangladesh. Hence, the signing of the cooperative arrangements to share waters from the transboundary rivers do not guarantee the end of water disputes.

Second, the roots of most of the transboundary water-related disputes in South Asia lie in the Partition of British India in 1947. Territorial division of the British India between India, Pakistan, and Bangladesh (then East Pakistan) disrupted most of the interlinked and interdependent water infrastructures the British imperialists built for their benefits. The Partition of British India witnessed a lot of violence in which many people lost their lives and a number of women were raped. Memories of such violence have embittered relationships between India and Pakistan since 1947, and, largely, guide them in making decisions about each other.

Third, unlike Pakistan and Bangladesh, other South Asian riparian neighbours of India, Nepal, and Bhutan express concerns over the Indian hydropower projects in their respective countries. There is a feeling gripping across a significant section of the population from these countries that India "exploits" their water resources for its own benefit. It persists, mainly, because of policies of the Indian HEP construction companies in those countries and the growing hydro debts, mainly, in Bhutan.

Structure of the book

Besides the introduction and the conclusion, this book contains five chapters. The sequence of chapters reflects India's bilateral relationships with the respective South Asian countries. They appear in descending order based on the degree of disputes and differences India has with its respective South Asian riparian neighbours.

The first chapter is "South Asia: region, history, and politics". This chapter looks at South Asia as a region and the history of bilateral relations and differences with the respective riparian neighbours are discussed. It also, briefly, talks about why China, which also shares waters with India, has not been included in this study.

The second chapter is "River water disputes between India and Pakistan". In this chapter, the irrigation system developed during colonial period and their disruption during territorial division in 1947 are discussed. Why and how IWT was signed between India and Pakistan is looked at. This chapter then discusses the water-related differences and disputes between India and Pakistan even after signing the IWT. It also looks at how and why transboundary rivers between India and Pakistan have become political.

The third chapter is "River water disputes between India and Bangladesh". The disruption of the canal system laid down by the British colonialists during partition of India and related difficulties are discussed. Controversy over the building of the Farakka barrage and its history is explored. The Ganges water treaty and differences over Teesta River water are discussed. This chapter also examines the impact of bilateral relationships between India and Bangladesh on their transboundary water issues. In this chapter, at many places, Calcutta, instead of Kolkata, has been used as the period discussed is before the name of the city was changed in 2001.

The fourth chapter is "River water and Hydroelectric Power Project issues between India and Nepal". Nepal's potential to generate hydropower and the significance of water in its economy is discussed. The Mahakali river water treaty, Gandaki River water treaty, and Kosi River water treaty are discussed. India's assistance in Nepal's HEP projects and its social and political impact are looked at.

Finally, the fifth chapter examines "Concerns over India-assisted Hydroelectric Power Projects in Bhutan". Bhutan's potential to produce hydropower and the engagement of Indian companies are looked at. The growing hydro debt in Bhutan and accusations against India for exploitation of Bhutan's HEP for its own benefits are examined.

The concluding chapter examines the hypothesis of the book and reviews some of the key assessments and arguments. It overviews the challenges and possible future trajectories.

Notes

1 United Nations Development Programme, Human Development Report. (2006). *Beyond Scarcity: Power, Poverty and the Global Water Crisis*. New York: UNDP, p. 210. http://hdr.undp.org/sites/default/files/reports/267/hdr06-complete.pdf. Accessed on 12 July 2018.
2 Ibid.

3 According to World Commission on Dams the number is 261. See *The Report of The World Commission on Dams. (2000). *Dams and Development: A New Frame Work for Decision.* London: Earthscan Publications Limited.

4 Wolf, Aron T. (1995). *Hydro Politics Along the Jordan River: Scarce Water and Its Impact on the Arab Israeli Conflict.* Tokyo, New York and Paris: United Nations University Press; Swain, Ashok. (2008). "Mission Not Yet Accomplished: Managing Water Resources in the Nile River Basin". *Journal of International Affairs,* Spring/Summer, Vol. 61, No. 2, pp. 201–213.

5 United Nations Department of Economic and Social Affairs. "International Decade for Action 'Water for Life' 2005–2015". www.un.org/waterforlifedecade/transboundary_waters.shtml. Accessed on 19 August 2018.

6 Ibid.

7 Ibid.

8 United Nations World Water Development Report. (2018). http://unesdoc.unesco.org/images/0026/002614/261424e.pdf. Accessed on 12 May 2018.

9 Ibid.

10 Cited in United Nations World Water Development Report. (2018). http://unesdoc.unesco.org/images/0026/002614/261424e.pdf. Accessed on 12 May 2018, p. 10.

11 Ibid.

12 Ibid., p. 12.

13 Falkenmark, Malin, Jan Lundqvist and Carl Widstrand. (1989, 11 January). "Macro-scale Water Scarcity Requires Micro-scale Approaches". *Natural Resources Forum,* Vol. 3, No. 4, pp. 258–267.

14 United Nations Development Programme, Human Development Report. (2006). *Beyond Scarcity: Power, Poverty and the Global Water Crisis.* New York: UNDP, p. 12. http://hdr.undp.org/sites/default/files/reports/267/hdr06-complete.pdf. Accessed on 12 July 2018.

15 United Nations Secretary General "Remarks at 2019 Climate Action Summit". https://www.un.org/sg/en/content/sg/speeches/2019-09-23/remarks-2019-climate-action-summit. Accessed on 29 October 2019.

16 Brown, Lindsay. (2019, 3 May). "Climate Emergency: What Is Climate Emergency", *BBC.* https://www.bbc.com/news/newsbeat-47570654. Accessed on 5 May 2019.

17 International Panel on Climate Change Report 2018. "Special Report: Global Warming of 1.5°C". https://www.ipcc.ch/sr15/. Accessed on 28 June 2019.

18 Chinese dam breach caused northeast floods: AFP. Rediff.com 2000, 10 July http://www.rediff.com/news/2000/jul/10china.htm. Accessed on 18 December 2018.

19 "Pakistan Writes to UNSC, Accuses India of Threatening Regional Security", Money Control, 23 February 2019. https://www.moneycontrol.com/news/india/pakistan-writes-to-unsc-accuses-india-of-threatening-regional-security-3574561.html. Accessed on 24 February 2019.

20 Cech, Thomas V. (2010). *Principles of Water Resources: History, Development, Management and Policy.* Hoboken, NJ: John Wiley & Sons.

21 Ibid.

22 Ibid.

23 Siddiqi, Toufiq A. and Shirin-Tahir-Kheli. (2004). Water Conflicts in South Asia: Managing Water Resource Disputes Within and Between Countries of the Region, Project Implemented by Global Environment and Energy in the 21st Century (GEE-21) and the School of Advanced International Studies, John Hopkins University (SAIS), Sponsored by the Carnegie Corporation of New York.

24 Ibid.

25 Ibid.
26 Ibid.
27 Ibid.
28 Ibid.
29 Ibid.
30 Ibid.
31 United Nations. (2013). "Convention on the Protection and Use of Transboundary Watercourses and International Lakes", pp. 6–15. https://www.unece.org/fileadmin/DAM/env/water/publications/WAT_Text/ECE_MP.WAT_41.pdf. Accessed on 15 February 2019.
32 Ibid.
33 Ibid.
34 Salman M. A. Salman. (2007, December). "The Helsinki Rules, the UN Watercourses Convention and the Berlin Rules: Perspectives on International Water Law", *Water Resources Development*, Vol. 23, No. 4, pp. 625–640. https://www.internationalwaterlaw.org/bibliography/articles/general/Salman-BerlinRules.pdf. Accessed on 2 March 2018.
35 International Law Association Berlin Conference. (2004). "Water Resources Law". https://www.unece.org/fileadmin/DAM/env/water/meetings/legal_board/2010/annexes_groundwater_paper/Annex_IV_Berlin_Rules_on_Water_Resources_ILA.pdf. Accessed on 18 July 2018.
36 Ibid.
37 Ibid.
38 Ibid.
39 International Law Association Berlin Conference. (2004). "Water Resources Law". http://www.cawater-info.net/library/eng/l/berlin_rules.pdf. Accessed on 12 June 2018.
40 According to Berlin Rules while determining equitable and reasonable use some factors, though not limited, taken into consideration are (written in verbatim): Geographic, hydrographic, hydrological, hydrogeological, climatic, ecological, and other natural features; b. The social and economic needs of the basin States concerned; c. The population dependent on the waters of the international drainage basin in each basin State; d. The effects of the use or uses of the waters of the international drainage basin in one basin State upon other basin States; e. Existing and potential uses of the waters of the international drainage basin; f. Conservation, protection, development, and economy of use of the water resources of the international drainage basin and the costs of measures taken to achieve these purposes; g. The availability of alternatives, of comparable value, to the particular planned or existing use; h. The sustainability of proposed or existing uses; and i. The minimization of environmental harm.
   The Berlin rules say that "weight of each factor is to be determined by its importance in comparison with other relevant factors. In determining what is a reasonable and equitable use, all relevant factors are to be considered together and a conclusion reached on the basis of the whole." International Law Association Berlin Conference (2004) "Water Resources Law" http://www.cawater-info.net/library/eng/l/berlin_rules.pdf. Accessed on 12 June 2018.
41 "Convention on the Law of the Non-navigational Uses of International Watercourses 1997". United Nations General Assembly. http://legal.un.org/ilc/texts/instruments/english/conventions/8_3_1997.pdf. Accessed on 12 August 2019.
42 "International Decade for Action 'Water for Life' 2005–2015". United Nations Department of Economic and Social Affairs. www.un.org/waterforlifedecade/transboundary_waters.shtml.

43 See Crow, Ben and Nivikar Singh. (2008). "The Management of Inter-State Rivers as Demands Grow and Supplies Tightens: India, China, Nepal, Pakistan, Bangladesh. www.researchgate.net/publication/23779638_The_management_of_inter-state_rivers_as_demands_grow_and_supplies_tighten_India_China_Nepal_Pakistan_Bangladesh/download.

44 Zhang, Hongzhou and Mingjiang Li. (2018). "China and Global Water Governance: New Devlopments". In Zhang, Hongzhou and Mingjiang Li (ed) *China and Transboundary Water Politics in Asia*. Oxon: Routledge, pp. 219–236.

45 Ibid.

46 Cited in Siddiqui, Shawahiq. (2013). "Engaging with the Global: Prospects for the 1997 UN Watercourse Convention Being Adopted in the Ganga Region". Issue Brief *The Asia Foundation..* https://asiafoundation.org/resources/pdfs/ORFIssuebrief64ShawahiqSiddiquiformail.pdf. Accessed on 10 June 2018.

47 Ibid.

48 Connell, Daniel. (2013, 10 April). "Water Wars, Maybe, But Who Is the Enemy?" Global Water Forum. www.globalwaterforum.org/2013/04/10/water-wars-maybe-but-who-is-the-enemy/. Accessed on 12 December 2018.

49 Lhuisset, Emeric. (2016). *Last Water War, Ruins of a Future*. Paris: Andere Ferre.

50 Glieck, Peter (ed). (2006). *The World's Water 2006–2007*. Chicago: Island Press.

51 Ibid.

52 Daoudy, Marwa. (2008). "Israel, Syria and Golan Heights". *Journal of International Affairs*, Spring/Summer, Vol. 61, No. 2, pp. 215–234.

53 "India Will Stop Its Share of Indus Water to Pakistan, Says Nitin Gadkari". (2019, February 21). *Business Today*. https://www.businesstoday.in/current/economy-politics/india-will-stop-its-share-of-indus-water-to-pakistan-says-nitin-gadkari/story/321030.html. Accessed on 18 July 2019.

54 United Nations Development Programme, Human Development Report. (2006). *Beyond Scarcity: Power, Poverty and the Global Water Crisis*. New York: UNDP, p. 223.

55 Waltz, Kenneth. (1979). *Theory of International Politics*. Reading, MA: Addison-Wesley.

56 Sagan, Scot D. and Keneth N. Waltz. (1995). *The Spread of Nuclear Weapons: A Debate*. New York: W.W. Norton.

57 Grieco, Joseph M. (1990). *Cooperation Among Nations: Europe, America, and Non-Tariff Barriers to Trade*. Cornell: Cornell University Press.

58 Keohane, Robert O. (1984). *After Hegemony: Cooperation and Discord in the World Political Economy*. Princeton: Princeton University Press, pp. 51–52.

59 Ibid.

60 Mitarny, David. (1966). *A Working Peace System*. Chicago: Quadrangle Books, pp. 26–27.

61 Wendt, Alexander (1992) "Anarchy is what States Make of it: The Social Construction of Power Politics " *International Organsation* 46 (2) p 391–425.

62 See Davenport, Andrew (2011) Marxism in IR: Condemned to a Realist fate?" *European Journal of International Relations* 19(1) 27–48.

63 Rasul Baksh Rais. 'Afghanistan: A Weak State in Path of Power Rivalries'. In Paul, T.V. (ed)(2011) *South Asia' Weak States :Understanding The Regional Insecurity Predicament*; New York: Oxford University Press. pp. 195–219

64 Zeitoun, Mark. (2012). *Power and Water in the Middle East: The Hidden Politics of the Palestinian-Israeli Water Conflict*. London and New York: I.B. Tauris.

65 Lowi, Miriam R. (1993). *Water and Power: The Politics of a Scarce Resource in the Jordan Basin*. Cambridge: Cambridge University Press.
66 Zeitoun, Mark. (2012). *Power and Water in the Middle East: The Hidden Politics of the Palestinian-Israeli Water Conflict*. London and New York: I.B. Tauris, p. 52.
67 United Nations Department of Economic and Social Affairs. "International Decade for Action 'Water for Life' 2005–2015". www.un.org/waterforlifedecade/transboundary_waters.shtml.
68 Wolf, Aron T. (2008). "Healing the Enlightenment Rift: Rationality, Spirituality and Shared Waters". *Journal of International Affairs*, Vol. 61, No. 2, pp. 51–73.
69 Ibid.
70 Iyer, Ramaswamy R. (2015). *Living Rivers, Dying Rivers*. New Delhi: Oxford University Press, p. 447.
71 "Blood and Water Can't Flow Simultaneously: PM Narendra Modi Gets Tough on Indus Treaty". (2016, September 27). *The Times of India*. https://timesofindia.indiatimes.com/india/Blood-and-water-cant-flow-together-PM-Narendra-Modi-gets-tough-on-Indus-treaty/articleshow/54534135.cms. Accessed on 17 November 2018.
72 "India Will Stop Its Share of Indus Water to Pakistan, Says Nitin Gadkari". (2019, February 21). *Business Today*. https://www.businesstoday.in/current/economy-politics/india-will-stop-its-share-of-indus-water-to-pakistan-says-nitin-gadkari/story/321030.html. Accessed on 23 February 2019.
73 Constitution of India, National Portal of India. www.india.gov.in/sites/upload_files/npi/files/coi_part_full.pdf, p. 155.
74 Fifteenth SAARC Summit Colombo, 2–3 August 2008 Declaration Partnership for Growth for Our People. http://saarc-sec.org/uploads/digital_library_document/15_-_Colombo,_15th_Summit_2-3_August_2008_-_for_printing.pdf. Accessed on 18 October 2017.
75 Ministry of Power, Government of India. "Guidelines on Cross Border Trade of Electricity". https://powermin.nic.in/sites/default/files/webform/notices/Guidelines_for_Cross_Boarder_Trade.pdf. Accessed on 27 March 2018, p. 3.
76 Deepak Adhikari. (2019, 20 September). "Nepal Power Export Plans in Doubt as India Reviews Options". *Nikkei Asian Review*. https://asia.nikkei.com/Politics/International-relations/Nepal-power-export-plans-in-doubt-as-India-reviews-options. Accessed on 12 January 2020.
77 Government of India, Ministry of Power "Guidelines for Import/Export (Cross Border) of Electricity-2018-regarding". Office Memorandum. https://powermin.nic.in/sites/default/files/uploads/Guidelines_for_ImportExport_Cross%20Border_of_Electricity_2018.pdf. Accessed on 25 January 2019.
78 Ibid.
79 Chellaney, Brahama. (2013). *Water: Asia's New Battleground*. Georgetown: Georgetown University Press.
80 Wirsing, Robert G., Daniel C. Stoll and Christopher Jasparro. (2013). *International Conflict Over Water Resources in Himalayan Asia*. London: Palgrave Macmillan.
81 Sinha, Uttam Kumar. (2016). *Riverine Neighbourhood: Hydro-politics in South Asia*. New Delhi: Pentagon Press.
82 Ranjan, Amit. (2020). *Water in Himalayan South Asia: Internal Challenges, Disputes and Transboundary Tensions*. Singapore: Palgrave.
83 Murada, Vishal. (2017). *India's Water Security Challenges: Myths, Reality and Measures*. New Delhi: Viz Publications.

84 Iyer, Ramaswamy. (2002). "Three River Waters Treaty". In Sahadevan, P. (ed), *Conflicts and Peacemaking in South Asia*. New Delhi: Lancers Publication, pp. 365–395.
85 Water Conflicts in South Asia: India's Transboundary Water Conflicts with Pakistan, Bangladesh, and Nepal, *BISS Journal* (Bangladesh Institute for Strategic Studies, Dhaka) Volume 36 Issue 1 January 2015, pp. 36–57.

1 South Asia

Region, history, and politics

Defining and recognising a region is a challenging task. Scholars and political commentators have kept on defining and redefining a region according to their political and economic comforts. Highlighting the difficulty of having a definition of "region", Hagerty and Hagerty (2005) finds that theoretically, everyone has their own vague idea about region whose construction and deconstruction, in reality, a much more volatile affair. For Hagerty and Hagerty, regions are ephemeral and intellectual constructs which keep on changing due to various factors such as technological developments, geopolitical events, demographic flows, scholarly fads and the numerous other dynamics that taken together, constitute human history.[1] This challenge aptly explains the situation in South Asia which, over the centuries and decades, has passed through a number definition, construction, and reconstructions.

Whatever the current definition or imagination of the South Asia as a region, it has its genesis during the colonial India when the British established their imperial rule in the region. They set up a single authority and united it into a single territorial and political unit. Before the advent of East India Company, Emperor Ashoka (268–232 BCE) and a few Mughal rulers such as Akbar (1542–1605), Jehangir (1569–1627), Shahjahan (1592–1666), and Aurangzeb (1618–1707), more or less, united such a vast or a large part of the territory under a single authority. During colonial rule, the present India, Pakistan, and Bangladesh were part of British India. Afghanistan (became independent in 1919), Nepal, and Bhutan were indirectly part of British rule. Sri Lanka was under it till 1948 and Maldives, formally, attained independence from the British in 1965.

After the 1857 rebellion of soldiers, the British Crown officially took over rule of India from the East India Company. British India was divided into two major regions-areas which were directly held by the British, and the princely States ruled by the British Crown-friendly Indian princes. In 1947, the British India was partitioned between India and Pakistan while the then princely States had options to join either India or Pakistan. Almost all of those falling in the Indian territory joined India and those in Pakistan joined Pakistan. However, the border States such as Junagadh and Jammu & Kashmir (J&K) remained indecisive till the last days of British rule in India. Junagadh had Hindu majority population under the Muslim prince, Muhammad Mahabat Khan III, who wanted to join Pakistan but eventually the State became a part of India and the prince and most of the other Muslim power elites of Junagadh fled to Pakistan. J&K was a Muslim majority State with a Hindu king. In October 1947, after the tribal invasion backed by the Pakistani army, prince Hari Singh of J&K signed a treaty of accession with India. After the first Kashmir War (1947–48), two thirds territory of J&K came under India and Pakistan got control over a part of J&K while a small portion remains with China.

Different from J&K and Junagadh was the case of Hyderabad. Unlike Junagadh and the J&K, Hyderabad was not a border State between India and then newly created Pakistan. In Hyderabad, Muslims were in majority in the city while in rural areas it was Hindus who are more in number. As the Nizam of Hyderabad, Mir Osman Ali Khan wanted to remain either independent or join Pakistan, violence broke out between Nizam's private militia called *Razakars* and those who wanted to join India. To control the situation, the Government of India launched "Operation Polo" on 13 September 1948 against Nizam and forced his forces to surrender on 18 September 1948.[2]

At the international level, after the Partition, India emerged as successor State of British India. In 1947, it was thought that the Partition of British India settled all communal differences, disputes, and demands between the Hindus and Muslims, but it has not. Even after 71 years, Partition evokes political sentiments in India and Pakistan. Also, institutional memories of Partition-related communal violence is the major reason for enduring tensions between India and Pakistan.

Although it emerged on the basis of religion, Pakistan could not remain united for a long period, and in 1971 East Pakistan was liberated with political and military help from India. Initially, India and Bangladesh shared cordial relationships, but after the assassination of Sheikh Mujibur Rahman in August 1975, differences began to emerge between the two countries. It improved after Sheikh Hasina came into power in Bangladesh in 1996. Since 2008, Hasina has been in power.

India and Nepal share the same religious majority but differences between the two States remain over a number of political issues. People from India and Nepal have more close relations than between the two governments. Bhutan is regarded as the politically closest neighbour of India.

This chapter examines the bilateral political relationships India has with its respective South Asian riparian neighbours. This will help to understand the nature and reasons for water disputes between India and its riparian South Asian neighbours.

South Asia – a region

As a political construct, the member countries of South Asia Association for Regional Cooperation (SAARC) are considered as a part of South Asia. This is not only a political definition of the region, but social and cultural factors too have an important role in shaping it. The South Asian countries which share borders with India also share social linkages and cultural bonding with the people living on the borderlands and across the border. For example, Indian Tamils and Sri Lankan Tamils belong to same ethnic group, people from West Bengal and Bangladesh share cultural similarities and language, the Indian Punjab and Pakistani Punjab share language and other cultural traits, and marriages between people from India and Nepal is a common practice, especially from border regions. Unlike them, the other countries may share a border with one or the other South Asian countries, but none form any such close bonding. Myanmar is an exception. It was part of British India till 1935. Myanmar shares a border and culture with India and Bangladesh but is still not considered as a part of political South Asia but of South East Asia.

After the end of the Cold War (1948–1991) in 1991, in the age of globalisation, a new definition of region has emerged. Countries have tried to increase connectivity and find market so that they can reap economic gains. This speed of connectivity has been halted with rise of protectionism, which is entirely opposite to the values of globalisation. In the age of globalisation and race for market, concept of "extended neighbourhood" has emerged. For India this "extended neighbourhood" stretches from the Suez Canal to the South China Sea, and includes within it West Asia, the Gulf, Central Asia, South East Asia, East Asia, the Asia Pacific, and the Indian Ocean Region.[3] Although this concept of "extended neighbourhood" is new to meet economic realities of the globalised world, it has been practised by the Indian security establishment since British India days and was inherited by independent India in 1947 from its erstwhile imperial masters. In the immediate neighbourhood, since 1947, as Priya Chacko (2012) finds, India has formulated its own Monroe Doctrine,[4] and has enacted policies to maintain its hegemony, as the British did during colonial rule in South Asia.

Other than bonding between the neighbouring countries and pursuing policies to set up an "extended neighbourhood", extra-regional actors too play a crucial role in South Asia and have relevant influence on securitisation or de-securitisation of the region.[5] This is not a new phenomenon, but rather has been taking place since the British colonial days. During the colonial period, the British India government used forward policy to check all such external interferences. During the Cold War days the former Soviet Union and the USA have interfered a lot in the matters related to South Asian politics. Virtually, through interferences they shape up and define the region.[6] At that time, Pakistan became an ally of the USA while India tilted towards the former Soviet Union. After India's defeat in the 1962 Sino-India war smaller South Asian countries and Pakistan established close relations with China and invited it in the region, largely, to check India. Consequently, after the end of Cold War in 1991, China gradually spread its wings to fill the vacuum created by the former Soviet Union. After 1991, from being an ally of the USA during the second half of the Cold War, China turned into its global competitor.

Geographical contiguity, necessities of globalisation, growing connectivity, and the role of China as an external power balancer in South Asia have made many South Asian countries such as Bangladesh in 2007 support its candidature into the SAARC. However, China does not have socio-cultural similarities with a large part of South Asia. Hence, it is not suitable to be a part of South Asia region. For all social, cultural, and political purposes, South Asia constitutes India, Pakistan, Bangladesh, Afghanistan, Nepal, Bhutan, Sri Lanka, and Maldives.

India-Pakistan relationships

In 1947 India and Pakistan became two sovereign countries after British India was partitioned. Although political differences between the Indian National Congress and the All India Muslim League were reasons for the Partition, there were many other complex factors which created an environment for it. One such factor was the lack of communal trust and a growing sense of alienation among the sizeable Muslim population. In some parts of British India, the Hindus were comparatively rich and were accused of practicing discriminations against the Muslims which became one of the reasons for alienation of a number of Muslims and acted as a catalyst for the Partition of British India in 1947.[7] Another significant factor was after the end of the Mughal rule in 1857, the erstwhile Muslim landlords and Muslim elites lost their socio-political status. In the hope of regaining such status many Muslim elites supported the Muslim League which during the last decade of the British rule called on for a separate State for the Muslims of British India.

Once the Partition was accepted, communal violence began in which a large number of people were killed, and many women were raped and abducted.[8] Although created as a country for the Muslims from British India, several Muslims preferred to stay back in India and likewise a small number of Hindus remained in the then newly constituted Pakistan.

Pre-partition differences, distrust, and institutional memories of Partition-related violence, still dictate the India-Pakistan relationship. Since 1947 the two countries have fought three total wars (1947–48, 1965, and 1971), one limited war (Kargil 1999), witnessed a military tensed situation (1987, 2002, 2008, 2016, and 2019), and always remain in political-cum-military tensions.[9]

Both India and Pakistan accuse each other of inciting tensions between their people and carrying out violent attacks in their respective territories. After eruption of militancy in the Kashmir valley in the late 1980s many Pakistan trained militants have intruded in the Indian J&K to fight Pakistan's proxy war against India.[10] Besides, J&K, in the past, Pakistan had been also engaged in fanning militancy and insurgency in India's northeastern States.[11] Pakistan's security agencies helped the leaders from the various insurgent groups in the northeast in their fight against the Indian state.[12] Then, in the late 1970s and 1980s, Pakistan also provided support to the Sikh militants. As claimed by the Indian agencies, Pakistan still provides support to individuals such as Manjit Singh, leader of *Khalistan Zinadabad Force*.[13] Pakistan-based militant groups have also carried out militant acts in parts of India such as an attack on Indian parliament in 2001, mayhem in Mumbai in 2008, etc. Militants operating from Pakistani soil, as the government of India maintains, have been actively engaged in carrying out attacks on the Indian security forces and the civilians.[14]

Pakistan counters all such charges and, instead, accuses India of fomenting internal tensions and blames the Indian agencies for inciting violence in the country. In 2017, Pakistan's security agencies arrested Kulbhushan Jadhav from Balochistan province. Pakistan argues that Jadhav is an Indian spy who was directing subversive activities in Karachi and Balochistan.[15] On contrary, India maintained that after his "premature retirement" from the Indian navy[16] the government has no links with him. Later, India approached to the International Court of Justice on the matters. The Court's advice on the issue has not brought respite to Jadhav, as he is still in prison in Pakistan.

On its relationships with Pakistan, the Ministry of External Affairs of India annual report of 2017–18 says that

> Our [India's] efforts to normalize relationships with Pakistan continued to be thwarted by Pakistan's support for cross-border infiltration and ceasefire violations along the Line of Control and International Boundary.[17]

The author could not find any such report on the website of Pakistan's Ministry of Foreign Affairs, but it has repeatedly blamed India for pulling out from the bilateral talks.[18]

It is not that India and Pakistan did not try to improve their bilateral relationships; they have made many attempts to do so but could not succeed. One of the serious attempts to improve India-Pakistan relationships by addressing all their issues was made in 1998 when the two countries decided to hold Composite Dialogues (CD). All of their bilateral issues were clubbed, and discussions were held. However, the spirit of CD died soon due to the Kargil (1999) war and other terror incidents in India. In 2004 discussions on CD resumed. One round of talks was held in 2004 and in

2005 they held the second round. Foreign secretaries from the two countries led the talks which was monitored by the foreign ministers from the respective countries.[19] The 2008 Mumbai terror attack stalled the talks. In 2010 the two countries re-engaged under the title "resumed dialogue" instead of CD.[20] Few issues such as counter terrorism, including Mumbai attack trials and humanitarian issues, which was to be overseen by the home secretary, were included in it.[21] Again in 2012, talks were stopped after an Indian soldier was beheaded by the Pakistan Army at the India-Pakistan border.[22] Since then the political leaderships from the two countries have talked to each other on some occasions such as the Indian Prime Minister Narendra Modi's landing in Lahore in December 2015, etc. but never engaged in CD or "resumed dialogue". In 2016 the two countries agreed on diplomatic talks but it was suspended after the Indian Air Force base in Punjab was attacked by the militants.[23] For any resumption of bilateral talks, India has repeatedly said that Pakistan has to create suitable environment as "terror and talks can't go together".[24]

After taking his office as Prime Minister of Pakistan, in July 2018, Imran Khan wrote a letter to Narendra Modi seeking to start again the bilateral talks to resolve bilateral issues for which he was slammed by the opposition parties from his country.[25] However, he has failed to prepare suitable environment for such engagements, as militant activities against India from Pakistan soil continue. Khan changed his earlier position after August 2019, when India changed the status of its side of J&K. First Pakistan said no talks and then in September Shah Mahmood Qureshi, foreign minister of Pakistan, said that it is ready to have "conditional" talks.[26]

In the absence of talks almost all issues between India and Pakistan remain unresolved. Major issues on which differences, tensions, and disputes between them exists are:

(a) *Siachen Glacier:* Siachen glacier is one of the most inhospitable and glaciated regions in the world. It receives about 10 meters of snow every year. Blizzards can be of a speed up to 150 knots (nearly 300 kilometres per hour). The temperature drops routinely to 40 degrees below zero on the centigrade scale.[27] In such inhospitable and almost inhabitable land, India and Pakistan maintain their troops at much higher economic and physical costs. The roots of this dispute lie in the United Nations mediated ceasefire agreement between India and Pakistan in 1949. It delineated the Line of Control until point NJ 9842, after which, it said it would run "thence north to the glaciers".[28] In 1984, fearful of adverse Pakistani military's moves in Siachen, Indian soldiers moved north and eventually occupied the highest points on the glaciers. This started the conflict over the Siachen glacier.[29] Claims from India and Pakistan over Siachen is, mainly, due to their interpretations of the vague language contained in the documents of 1949 and 1972 Shimla agreements.

Pakistan draws a straight line in a northeast direction from NJ 9842 to the Karakorum pass on its boundary with China while India draws a north-west line from NJ 9842 along the watershed line of the Saltoro Range, a southern offshoot of the Karakorum Range.[30] To address the Siachen conflict, at a meeting between the then Indian Prime Minister Rajiv Gandhi and then Pakistan's President General Zia-ul-Haq, on 17 December 1985, an agreement was reached to hold defence secretary level talks. The first round of talks was held between 10 and 12 January 1986 at Rawalpindi.[31] After a series of talks, in 1992, India and Pakistan had reached a near agreement on the Siachen after Islamabad gave its assent to record the then existing number of troop positions in an annex of the agreement. However, the deal could not finalise because the Indian political leadership developed cold feet over it. Earlier, Pakistan's proposal in an annexure on vacation of areas found an immediate acceptance among the Indian officials.[32] The Indian delegation was led by N.N. Vohra, then defence secretary. On the failure of Siachen deal, N.N. Vohra said to a reporter from *The Hindu* that:

> We had finalized the text of an agreement at Hyderabad House by around 10 pm on the last day. . . . Signing was set for 10 am. But later that night, instructions were given to me not to go ahead next day but to conclude matters in our next round of talks in Islamabad in January 1993. . . . Of course, that day never came . . . that's the way these things go.[33]

> Siachen is still a militarised glacier where many soldiers from India and Pakistan lose their lives because of cold, avalanches, and other natural accidents. In 2012, 140 people, including 129 Pakistani soldiers, lost their lives due to avalanche. In 2017, 24 people including 20 lost their lives. In 2019, four Indian Army soldiers and two civilian porters lost their lives, as they were hit by avalanche.[34]

(b) *Sir Creek Dispute:* The Sir Creek is a 96-kilometres-long estuary in the marshes of the Rann of the Kutch, which lies on the border between the Indian state of Gujarat and the Pakistani province of Sindh. This dispute has its origin in the compromise reached between the government of Sindh in British India and the Kutch Darbar (court) over boundary delimitation over the Kori Creek, which originally divided the two principalities in 1908, and lies east of the Sir Creek.[35] 1914 map shows a green line running along the eastern bank of the Sir Creek on the Kutch side of the river as the boundary. Pakistan's position is that the boundary is on the eastern bank, called green line, due to historical title seems accurate.[36] Sikander Ahmed Shah, former legal advisor to the government of Pakistan, in his 2015 article in *Dawn* further writes that "Under the established principle of uti possidetis [may you continue to possess such as you do possess] both Pakistan and India, on gaining independence, inherited the original borders of their predecessor state and the delimitations effected by them. In this regard, Resolution 1192 [1914] classifies the B-44 map [of 1914] as authoritative and expressly rejects the applicability of the thalweg principle".[37]

On contrary, the Indian view is that the green line mentioned in the 1914 map was symbolic. India claims Sir Creek is navigable in high tide, therefore boundary should be drawn from the mid-channel, while Pakistan argues it is not navigable.[38] India's claim is, mainly, based on another map drawn in 1925, and implemented by the installation of mid-channel pillars back in 1924.[39] To support its position India has also cited thalweg[40] principles.

In 1965 the tribunal, under a judge of a Swedish Court, Gunnar Lagergren, was set up to demarcate only the northern border of the Kutch-Sindh sector between India and Pakistan. The Sir Creek was a part of the territorial dispute but was left out of the tribunal's jurisdiction. The tribunal announced its final verdict on 19 February 1968 in Geneva[41] in which it allotted 90% of Kutch to India and 10% to Pakistan.[42] Later, both sides had resolved to settle the dispute over Sir Creek in a speedy manner, given their obligations under the UN Convention on the Law of the Sea (UNCLOS). Both India and Pakistan have signed and ratified the UNCLOS. Between 1969 and 2012 surveyors from both countries have held 12 rounds of talks but failed to decide over the demarcation of the marshy land.[43]

(c) *Terrorism*: Terrorism is one of the major irritants between India and Pakistan. The seeds were sown by Pakistan after the end of Afghanistan war of 1989, to bleed India,[44] but over a longer period, Pakistan has also faced the consequences. Earlier the Pakistan Army had trained, groomed, and regulated the terror groups[45] but, mainly, after Pakistan's decision to become a part of the USA-led Global War on Terror (GWOT), the groups began attacking their one-time supporter.[46] In those attacks many Pakistani citizens have lost their lives. With connivance of their sympathizers the militants have even attacked military installations. In June 2011, the army was forced to investigate Brigadier Ali Khan for his ties to the militants of Hizb-ul-Tahir, a radical organisation that seeks to establish a global caliphate and thinks that its mission should begin from nuclear Pakistan.[47] Terrorists have not even spared the school students. On 16 December 2014 Tehrik-i-Taliban Pakistan (TTP) attacked on Army Public School, Peshawar in which 144 students and staff members were killed.[48]

As terrorism turned menace for both countries, earlier, to fight together in 2006, on the sidelines of the Non-Alignment Movement's meeting, in Havana (Cuba), then Indian Prime Minister Dr Manmohan

Singh and then Head of Pakistan General Parvez Musharraf met and issued a joint statement for creating Joint Anti-Terror Institutional Mechanism (JATM) on 16 September 2006, in which both countries pledged to cooperate to deal with terrorism.[49] The first meeting of JATM was held at Islamabad in March 2007.[50] However, it never came into full-fledged operation, as trust-deficit between the institutions remains high.

The 2008 Mumbai attack carried out by a Pakistan based terror group had tensed the bilateral relations between India and Pakistan. India accused Pakistan of giving shelter to terrorists who planned the attack while Pakistan responded that India had not provided it with enough evidence. Those who planned and executed the Mumbai attack of 2008 are: Zaki-ur Rahman Lakhvi, Abdul Wajid, Mazhar Iqbal, Hamad Amin Sadiq, Shahid Jameel Riaz, Jamil Ahmed and Younis Anjum.[51]

Shockingly, Pakistani media accused the Indian intelligence agency of carrying out and supporting violence within its territory.[52] Pakistan also accused India of funding the Baloch nationalists from its Afghanistan based consulates at Herat, Mazar-i-Sharif, Kandahar, and Jalalabad.[53] According to a report published in the British Broadcasting Corporation (BBC), Muttahida Qaumi Movement (MQM) militants had been trained by India and the party has received funds for its activities. The Indian authorities have called it "completely baseless".[54] The report lacks substantive evidence to prove its contents, but in the past the Indian spy agency did carry out operations inside Pakistan to stop it from supporting Sikh militancy in India. It is public now that I.K. Gujral, after becoming India's Prime Minister, closed down all offensive operations in Pakistan, forcing the Research and Analysis Wing (RAW) to close down the CIT cells (J and X) that were used for them.[55]

Sometimes common people, especially those living near land border and fishermen fishing in international waters, are taken as militants and arrested for long time or get killed. One such incident was in 2014 when a fishing boat from Pakistan was blown up by Indian Coast Guards. Fishermen were reported a part of terror groups on a 2008 Mumbai like mission. Truth came in light when the then Deputy Inspector General of Coast Guard, B.K. Loshali, accepted that he ordered to "blow the boat off, we do not want to serve them biryani".[56]

Cross border terrorism remains a serious issue between the two countries.

(d) *Jammu & Kashmir:* Some political commentators consider J&K as a core[57] of India-Pakistan tensions. The dispute is more on the Kashmir valley and not so serious over on the status of Jammu and Ladakh. Both India and Pakistan claim J&K as their territory. For India, secularism is the guiding principle ; while for Pakistan, which was formed in the name of Islam, J&K (mainly Kashmir valley) is important to complete the promises made to it at the time of the partition of India. Due to this incompatible goal the conflict persists, and no party wants to compromise. Hence, the Kashmir dispute is more of an ideological, rather than a territorial one.[58] In the past, in 1954, 1963, and in 1972, there were moments when this conflict could have resolved but could not. Then in 2007 the two leaders were almost agreed on the draft over Kashmir issue, but, like past, it did not happen.[59]

Disputes over J&K has been complicated after the rise of the militancy in late 1980s. Its genesis can be found in the result of the 1987 assembly elections which produced disgruntled individuals who have led the militant and separatist movement in the Kashmir valley. In the elections, Muhammad Yusuf Shah was leader of the Muslim United Fronts (MUF), which was a coalition of Islamic parties.[60] Mohammad Yasin Malik, then 21-years-old, was his election manager. Voting was rigged and the popular will of Kashmiris was repressed in that election which convinced many of the MUF supporters that armed revolt is the only way forward to achieve one's rights.[61] Yusuf Shah, now known by his *nom de guerre*, Syed Salahuddin, has since the early 1990s been commander-in-chief of Hizb-ul-Mujahideen, the largest guerrilla force fighting in the Kashmir valley.[62] Yasin Malik leads the Jammu & Kashmir Liberation Front.

Post-1987, conducting elections in Kashmir valley have become practically a test for Indian democracy. In 2017, parliamentary by-elections in Srinagar constituency only 2% votes were polled. The election was deferred when eight people were killed and only 7% were polled due to simmering tensions, a rise in violence, and an increase in anger among the people of the valley against the Indian establishment.[63]

In 2019, the Hindu nationalist, Bhartiya Janata Party (BJP)-led National Democratic Alliance revoked the Special Status granted to J&K under the Indian constitution's Article 370. It was incorporated in Part XXI of the Indian Constitution in October 1949. The title of the part is "Temporary, Transitional and Special Provisions" under which special provisions given to many of the Indian States have been incorporated. Those who supports Article 370 argue that it was accepted by the Indian Constituent Assembly in 1949 and adopted by the J&K Constituent Assembly which was convened in 1951. Second, N. Gopalsawami Ayyanagar who drafted the article felt that for a variety of reasons, the situation in the J&K was not then ripe for integration. The major reason he cited for it was that a large chunk of the State was under control of the rebels and enemies.[64] Third, as also argued by the Pakistani commentators, there were United Nations Security Council resolutions of 1948 which called on for the Plebiscite in the region to ascertain the wishes of the people. Plebiscite was to take place after the Armies from the two countries leave J&K. However, it never happened.[65]

After revoking the Special provision, the Indian government has turned the entire J&K into a political cage. All communication lines have been curbed and people are living inside their home due to curfew. The mainstream political leaders have been detained and hundreds of youths are in various jails across India. The status of J&K has also been demoted from state to a Union Territory and it has been bifurcated into J&K and Laddakh. Although the government of India claims that situation is "normal" and the region will develop under the new arrangement, one has to observe the situation. Some opine contrarily and feel that the revocation of Special Status would cause further alienation among the people from the Kashmir valley.[66]

Pakistan reacted strongly against India's decision. Protests were held across Pakistan in solidarity with Kashmiris. It suspended the trains to India; the Pakistani High Commissioner to India-designate, Moin-Ul-Haque, was held back from assuming his charge; the Indian High Commissioner, Ajay Bisaria, was sent back to India; and trade with India was stopped.[67]

(e) *Economic and commercial cooperation*: In the past, India and Pakistan have tried to cooperate in economic sectors. It was, largely, believed that the economic cooperation could generate positive spill over to improve bilateral relations. In 1996, India granted Most Favoured Nation (MFN) status to Pakistan. Years later, in April 2012, Integrated Check Post (ICP) at the Wagah-Attari became operational. A study released by the Associated Chambers of Commerce and Industry of India in 2012 said that with the ICP becoming operational and if Pakistan granted the MFN status to India, the annual bilateral trade between the two countries could increase from US $2.6 to US $8.8 billion within next two years. The two countries decided to have 13 such trade Check Posts.[68] However, there were many impediments which they tried to cross but could not succeed. They could not implement the South Asian Free Trade Agreement and build consensus over the negative list. Even for so many years, largely due to political reasons, Pakistan had not granted MFN status to India.

During all those years, whatever trade relations the two countries had eventually collapsed after the militant attack on a Central Reserve Police Force convoy in Pulwama in Kashmir in 2019. After the attack, India withdrew MFN status from Pakistan. It also exorbitantly increased the duties on goods from Pakistan. Later in

August 2019, after India revoked the Special Provisions in Article 370 of the Indian constitution from J&K, Pakistan broke almost all remaining trade links with India except import of some of the very essential goods such as life-saving medicines.[69] The decision to stop trade has plunged India-Pakistan trade. In the first half of the 2019–20 fiscal years, Pakistan's exports to India came down to US $16.8 million compared to US $213 million in the first half of 2018–19. Imports from India to Pakistan also fell to US $286 million as against US $865 million in the first half of 2018–19. Hence, the trade deficit Pakistan has with India US $269.8 million.[70]

The water issue has not been mentioned here, as it was discussed in Chapter 2 of this book.

India-Bangladesh relationships

As India assisted in Bangladesh's liberation war of 1971, India and Bangladesh began their relationships cordially. After the liberation of the country, Sheikh Mujibur Rahman became the first President of Bangladesh. To deepen their friendship and address some of their problems, in 1972 India and Bangladesh signed a friendship agreement.

Not happy with Sheikh Mujib's leadership, a group of military men killed him and his family members who were present with him on 15 August 1975. After the assassination of Mujib, in quick succession, Khondaker Mostaq Ahmad (15 August 1975 to 6 November 1975) was appointed as the President of the country. He was succeeded by Abu Sadat Mohammad Sayam (November 1975 to April 1977).[71] These two were ceremonial heads; the real power was with the military. Later, in 1977, General Ziaur Rahman took over as Head of the State. Zia was assassinated in 1981. His policies were also followed by the H. M. Ershad (1983–1991) led military government. During the rule of the two military generals, India-Bangladesh relations lost the early days sheen. In 1991, democracy returned to Bangladesh. In that year's election, Khaleda Zia's Bangladesh Nationalist Party (BNP) came into power. Her coming into power did not substantially change the India-Bangladesh relationship. In the 1996 elections, Sheikh Hasina's Awami League (AL) won and she became the Prime Minister of Bangladesh. During her tenure India and Bangladesh signed the Ganges Water sharing treaty and tried to explore opportunities to create many bilateral engagements. In 2001, Khaleda Zia was once again back in power, and from 2006 to January 2009 the caretaker government run the country. In January 2007 Dr Fakharuddin Ahmed became chief advisor to that caretaker government. Elections were held in December 2008 and in January 2009, Sheikh Hasina came into power and she also won the subsequent elections in 2014 and then in 2018.

During Hasina's tenure, India and Bangladesh signed a protocol to implement Land Boundary Agreement (LBA) in 2011. This agreement was signed by Indira Gandhi, then Prime Minister of India and Sheikh Mujibur Rehman, then President of Bangladesh in 1974. The LBA came into effect in 2015 after the ratified agreement was exchanged between the two countries during Narendra Modi's visit to Dhaka in June 2015. In the land swapping exercise under the LBA, India has received 2777.038 acres of Adverse Possession and transferred 2267.682 acres of the same form of land to Bangladesh. In enclaves, India received 51 (7,110.2 acres) of the 71 Bangladeshi enclaves that are inside India proper. Bangladesh received 111 Indian enclaves (17,160.63 acres).[72]

Despite a close relationship, mainly, with Hasina at the helm of affairs in Bangladesh, the two countries have differences over a number of issues such as:

(a) *"Illegal" immigration and population movement across the India-Bangladesh border:* "Illegal" immigration of Bangladeshi citizens and population movement across the India-Bangladesh border is a cause for tensions between India and Bangladesh. Early evidence of large-scale immigration from East Bengal into Assam can be traced back to the late 1820s and 1830s when the tea plantations[73] started on a large scale. By the 1850s, this industry expanded and required a large number of workers.[74] To facilitate the process of bringing workers to work in the tea plantation sector, the colonial government made a series of legislations from 1863 to 1901.[75] A few years after the tea plantation sector developed, oil was detected in Assam. This sector too attracted many labourers from other parts of India, including Bengal.[76] In post-independent India, Assam witnessed the influx of a large number of refugees from (erstwhile East Bengal) East Pakistan after the Pakistani army started atrocities against the country's Bengali-speaking population in 1971. After its liberation, in 1972 India and Bangladesh signed an agreement under which the two countries agreed on 24 March 1972 as a cut-off date to determine citizens and refugee. This date had not been not accepted by many groups in the Indian state of Assam which called on to review it and wanted 1951 as a cut-off year. On this issue anti-foreigner sentiments rose in Assam and led to the Assam movement/Assam Agitation (1979–1985) under organisations such as the All Assam Students Union and the All Assam Gana Sangram Parishad. At that time, the Indian government enacted the Illegal Migrant (Determination by Tribunal) Act 1983.[77] Later, the Assam Accord was signed in 1985. It ended the movement. Not satisfied by the works of the IM(DT) Act's work, a group moved to the Supreme Court of India against the issue.

The problem of immigrants is not only in Assam but also in the Indian state of West Bengal. In 1999 the issue of taking back the people from

Bangladesh illegally living in India was raised by Jyoti Basu, then Chief Minister of West Bengal, with the Sheikh Hasina during her visit to Kolkata. Deb Mukherjee, the then Indian High Commissioner to Bangladesh, told Ranjan Basu of *Bangla Tribune* (also published in *Dhaka Tribune*) in 2018 that Jyoti Basu said West Bengal had to face economic pressure largely because most of the Bangladeshi intruders settle in West Bengal, and only a few of them migrate to Delhi and Mumbai. The intruders include both Hindus and Muslims. He demanded to Hasina that Dhaka must take back the "illegal Bangladeshis".[78] To it, Hasina said:

I do not believe a single Bangladeshi national is living in India illegally. If any of them does, then show me a list. You just cannot make a complaint. I want the India government to hand me a list of illegal Bangladeshis . . . Besides, the situation in Bangladesh is not so bad that people will cross the border and live in your country. And why would they do so?[79]

In 2005, the Indian Supreme Court struck down the IM (DT) Act. In 2015 a process to update the National Register of Citizens (NRC) for Assam began. The first draft was published at midnight of 31 December 2017.[80] On 30 July 2018, the final draft of the updated NRC was released at Guwahati. It declared more than 28.9 million out of about 32.9 million applicants from the Indian state of Assam as "eligible for [Indian] citizenship".[81] After the two drafts, the final list was published on 31 August 2019 in which 1.9 million people have been found not eligible for Indian citizenship.

After the NRC process began, Bangladesh expressed fear that India may push people into their country. To dispel such fears, then Indian High Commissioner to Bangladesh, Harshvardhan Shringla, made a statement at Dhaka that, "This [NRC] is entirely an internal matter of India".[82] *The Indian Express* reported,

Then, weeks before the publication of the final draft, India had quietly and informally briefed Bangladesh on the draft National Register of Citizens (NRC) in Assam, it is learnt, and assured them that there was no talk of "deportation" to prevent a slide in bilateral ties.[83]

During his visit to Dhaka on 13 July 2018, India's then Home Minister, Rajnath Singh, briefed Bangladesh Home Minister Asaduzzaman Khan on the "broad contours" of the NRC.[84] Following the release of the final draft, Shringla met Bangladesh Foreign minister A H Mahmood Ali and Awami League's general secretary and Transport Minister of the country, Obaidul Quade.[85] However, a number of Bangladeshis have been deported to Bangladesh in successive years. For example, in recent times, 5,234 "illegal migrants" were deported to Bangladesh in 2013 and 1,822 such deportations were accepted by the Bangladeshi government between 2014 and 2017.[86] In its 2014 judgment, the Supreme Court directed the Union government to "enter into necessary discussion with the Government of Bangladesh to streamline the procedure of deportation".[87]

In December 2019, the Citizenship Amendment Act (CAA) was passed by the Indian parliament. CAA has a provision to give citizenship to the persecuted Hindus, Sikhs, Parsees, Christians, and Jains from Bangladesh, Pakistan, and Afghanistan. They have to pass certain other required eligibility criteria such as duration of residence in India which is now "not less than five years".[88] This act came into effect in January 2020. Across India, CAA is being opposed because, as argued by the protestors, it tarnishes the secular character of India and, beyond that, it is hyphenated with the NRC[89] which the Indian Home Minister, Amit Shah, has repeatedly said will be carried out across the country.[90]

Both NRC and CAA are being strongly opposed in Bangladesh. After CAA came into effect, Narendra Modi was supposed to visit to Dhaka on March 17 to join the celebrations marking Sheikh Mujibur Rahman's birth centenary. This was protested by various outfits from Bangladesh and organised by Samamana Islami Dalgulo and Islami Andolan Bangladesh.[91] The day was saved as Bangladesh cancelled the celebration due to COVID-19 situation.

(b) *Growing Bangladesh-China engagements:* The foundation of the China-Bangladesh relationship was laid down by General Ziaur Rahman in 1977. In 2002, the Bangladesh Nationalist Party (BNP)-led coalition government in Bangladesh adopted a "Look East" policy to maximise economic and strategic achievements that could emerge from closer relations with East Asian countries, especially China. On Bangladesh's invitation China was added as an observer in the SAARC in 2007.

Trade between the two countries is expected to be around US $18 billion by 2021.[92] To boost their economic relationship, during the visit of the Chinese President Xi Jinping in October 2016, the two countries inked a total of 40 agreements and MoUs worth more than US $25 billion.[93] In his pre-visit speech, published in a Bangladeshi newspaper *The Daily Star*, Chinese President mentioned that

China is now the largest trading partner of Bangladesh and Bangladesh [is] China's third largest trading partner and third-largest project contract market in South Asia. Bilateral trade soared from USD $900 million in 2000 to US $14.7 billion in 2015, registering an annual increase of around 20%.[94]

More than trade, the growing defence partnership between China and Bangladesh worries India. In 2002, Bangladesh and China signed a defence cooperation deal.[95] Currently, Bangladeshi Navy is largely made up of Chinese-origin vessels.[96] Describing Bangladesh-China relations, in October 2016, after the signing of the MoUs and other Agreements, Xi Jinping said,

We agreed to elevate China-Bangladesh ties from a comprehensive partnership of cooperation to a strategic partnership of cooperation. We have agreed to increase high-level exchanges and strategic communication so that our bilateral relation will continue to move ahead at a higher level.[97]

(c) *Rise of militancy and religious radicalism in Bangladesh:* During General Zia's (1977–1981) tenure, a series of amendments were made in the Constitution to change the secular character adopted initially by Bangladesh. Under Zia the word "secularism" was deleted from the Preamble and Article 8 of the Constitution and a sentence "absolute trust and faith in the Almighty Allah" should be "the basis of all actions" was inserted. The words "Bismillah-ar-Rahman-ar-Rahim" (In the name of Allah, the Beneficent, the Merciful) were also inserted above the Preamble to the Constitution.[98] This Islamisation process was continued even by his successor, General Hussain Mohammad Ershad (1982 to 1990). In the 1980s, changes were visible not only in the politics of Bangladesh but also in the society which was passing through a social transformation process. A new version of Islam (mainly Wahhabism) silently replaced the centuries-old Sufistic version of Islam[99] practised by the Bangladeshis. Subsequent to widespread radical version of Islam, existing violence, deep political divides, and socio-political alienation of non-Bengali speakers together proved to be breeding ground to militancy in Bangladesh. Agent of radicalization and militancy have been, mainly, domestic actors, but involvement of outside actors could not be entirely ruled out, as many Bangladeshi workers living in West Asia have been influenced by the developments in the region or by the preaching of clerics there.[100]

In recent years, one of the most daring operations carried out by the militants in Bangladesh was on 1 July 2016 when seven fully armed militants stormed into Dhaka's popular eatery – Holey Artisan Bakery – and held about 60 hostages, mostly foreigners.[101]

In some cases, often the militant groups in Bangladesh commit their acts in their country and then cross into the Indian side of the border. Some of them have been found engaged in criminal activities in India. For example, in 2014, India's National Investigation Agency uncovered a suspected plot against Hasina when they captured members of

Jammat ul Mujahideen Bangladesh (JMB) in West Bengal.[102] There are Bangladesh-based groups such as JMB which, according to the Indian intelligence reports, looking for establishing base in West Bengal, Tripura, and Assam in India. The JMB have links with the Pakistan-based Lashkar-e-Tayebba (LeT). Modi government has designated it as a terrorist group in India.[103]

To fight against the militancy in Bangladesh and also to address its own insurgency-related concerns in northeast India; India has, after the LBA extended its support to Bangladesh. During his visit to Dhaka in May 2016, then Indian foreign Secretary S. Jaishankar conveyed India's support to Bangladesh in its fight against extremism and terrorism, particularly in response to attacks against vulnerable sections of society.[104] To fight against extremism and militancy, the two countries have been carrying out joint military exercises since 2010.

India-Nepal relationships

India and Nepal share about 1,800 kilometres of open and porous borders. Historically, the pre-2008, monarchy under the Shah Kings of Nepal (1768–2008) traces its roots to Sisodia Rajput of Rajasthan. In Nepal, they first established a state in the area marked as Gorkha under Drabya Shah in 1559.[105] After the East India Company (EIC) strengthened itself in India, it looked at Nepal to spread its influence. This led to the Anglo-Nepali War which went on for two years and ended with the imposition of the Sugauli Treaty on Nepal in 1816.[106] Once relations were established between Nepal and the EIC, Gurkhas became a part of the EIC army which they continued after the British Crown took over the imperial reign of India from them.

After the British left the Indian subcontinent in 1947, India and Nepal signed a friendship treaty in 1950. According to Article VI of the treaty

> Each government undertakes, in token of the neighbourly friendship between India and Nepal, to give to the nationals of the other, in its territory, national treatment with regard to participation in industrial and economic development of such territory and to the grant of concessions and contracts relating to such development.[107]

And Article VII says

> The Governments of India and Nepal agree to grant on a reciprocal basis to the nationals of one country in the territories of the other the same privileges in the matter of residence, ownership of property, participation in trade and commerce, movement and other privileges of a similar nature.[108]

Beyond the treaty between the two governments, there exist a strong age-old roti-beti[109] relationships, mainly between people from *Terai* part of Nepal and north Bihar.[110]

Despite all such arrangements, India-Nepal relationship was affected in the 1950s due to factionalism in the government of the day over issues like occupation of the post of Prime Minister and President of the Party by B.P. Koirala and Bharat Shamsher, leader of the Gorkha Parishad.[111] Also, at that time, India-Nepal friendship treaty was suspected by a section of Nepali who expressed their displeasure when the first Prime Minister of India Pandit Jawaharlal Nehru used the term "special relationship" to define India's relations with Nepal.[112] Some of the Nepali leaders were resentful, as they thought that under the grab of "special relationships" India may interfere in their sovereign affairs. Nehru's mistake was repeated in June 1969 by the then Minister of State for External Affairs Dinesh Singh during his visit to Kathmandu emphasised on "special relation" between the two countries.[113] The Chinese media picked it up and in its 28 June 1969 issue Chinese news agency *Hsinhua* called this "special relationship", as a part of India's expansionist policy.[114] Subsequently, protests were held against the presence of Indian check posts with armed guards in north-western border of Nepal.[115] Those protests took some days to silenced down.

During his tenure King Mahendra (1956-1972) took many decisions, such as allowing China and Pakistan to build their presence in order to counter what he found to be India's pressure for the restoration of democratic order in Nepal.[116] Not only the Chinese were allowed to build Kathmandu-Kodari Road but the Chinese traders and officials were also encouraged to make their presence in the *Terai*, proximate to India's Indo-Gangetic region through newly established chain of State Trading Corporation outlets in Nepal.[117] Then, to India's dismay, in 1988–90 King Birendra, successor of Mahendra, attempted to purchase Chinese arms which was perceived as a step to erode arrangements made in the India-Nepal friendship treaty of 1950. At that time, differences also surfaced between the two countries over trade and transit treaty and Nepal planned to ask Indians working in Nepal to have work permits. Nepal's growing closeness with China and differences over trade and transit angered India.[118] In reaction, India carried out an economic blockade of goods supplying via Indian territory into Nepal. Such a blockade remained for 13 months. The relationship between India and Nepal was normalised in 1990 after the removal of Panchayat form of governance (1960-1990) and restoration of democracy in Nepal. Then Prime Minister of Nepal Krishna Prasad Bhattarai visited New Delhi.[119]

Although a democratically elected government was in place in Nepal since 1990, the monarchy was a part of Nepal's political system and actively engaged in the country's affairs. The democratic voices gained overwhelming success in 2005 when after a second *janandolan* (people's movement) monarchy was abolished in Nepal. The 2005 movement was led by Seven Party Alliance.[120] After abolishing the monarchy, a second Constituent

Assembly (CA) was set up. During such a transition of political systems in Nepal, India played a significant role by maintaining distance from the palace. After the new government was formed under the Maoist leader Pushp Kamal Dahal, popularly called Prachanda, a fear was expressed about his policy towards India, but nothing of that sort happened. He followed the conventional approach during his first term (2008–09). He became Prime minister once again in 2016–17 and navigated cautiously between India and China.[121]

The CA set up in 2006 passed through various phases and came up with a final Constitution in 2015. Some of the provisions in the constitution, as alleged, were discriminatory towards many ethnic groups like Tharus, Madhes, etc. This stirred protests from Madhes which the Nepali leadership alleged supported by India. An early round of problems between India and Nepal after abolition of monarchy began after K.P. Sharma Oli became the Prime Minister of Nepal in 2015.[122] On constitution front, after a series of strong protests, the government of Nepal agreed to make a few amendments to make it fully inclusive. In support of agitators, India was accused by Nepal of carrying out an economic blockade from September 2015 to February 2016 which affected the supply of essential goods such as gas cylinder, petrol, etc., affecting the lives of common people.[123] However, the Government of India denies any such accusations and allegations.

In recent times, India-Nepal relations plummeted after India published an updated map depicting physical and political status of the two Union Territories – J &K and Ladakh. This map shows Pakistan-administered Jammu & Kashmir as a part of the Union Territory of Jammu & Kashmir while Gilgit-Baltistan (India calls it an occupied area by Pakistan) is placed under the Union territory of Laddakh.[124] Besides, the updated map also shows Kalapani in the Pithoragarh district of the Indian state of Uttarakhand. For Nepal, Kalapani is a part of its Darchula district in Sudurpaschim province.

Over the issue of showing Kalapani as an Indian territory, protests were mounted in Nepal against what they call they see and call "encroachment" or "occupation" of Nepali territory by India.[125] The Ministry of Foreign Affairs, Government of Nepal, press release in Nepali language saying that, "Nepal is clear about Kalapani being a part of Nepal's territory".[126] It adds "any unilateral decision regarding outstanding issues that need to be sorted out through mutual agreement will not be acceptable to the Nepal government".[127] The press release highlights that the two countries should resolve their border issues "diplomatically based on the historic documents and concrete evidence".[128]

India counters all such charges and maintains:

Our [India's] map accurately depicts the sovereign territory of India. The new map has in no manner revised our boundary with Nepal. The Boundary Delineation Exercise with Nepal is ongoing under the existing mechanism. We reiterate our commitment to find a solution through dialogue in the spirit of our closer friendly bilateral relations. At the same time, and I think it is very important to note that, both sides should guard against vested interests who are out there to create some differences between the two countries.[129]

India's stance on Nepal's affairs have made Nepal engage more with China. Nepal and China shares 1414 kilometres of border in the Himalayan range of the northern side of Nepal. Oli visited China in March 2016, during which both sides signed several different agreements and MoUs including the Transit and Transport Agreement. On 12 May 2017, Nepal and China signed a Memorandum of Understanding (MoU) on cooperation under the Belt and Road Initiative. This is expected to open several other new avenues in their bilateral cooperation such as economy, environment, technology, and culture. It also aims at promoting cooperation on policy exchanges, trade connectivity, financial integration, and further connect people.[130] After the devastating earthquake in Nepal in 2015 China provided assistance. Later, China provided 3 billion Yuan (US $42.3 million) on Nepal's Reconstruction to be used in the jointly selected 25 major projects for the 2016–18 period.[131] Then, on 23 December 2016, Nepal and the People's Republic of China signed an Agreement on Economic and Technical Cooperation in Beijing to provide grant assistance of RMB 1 billion (US $14 million) to the Government of Nepal for implementing the Syaphrubesi-Rasuwagadhi Highway Repair and Improvement Project, Upgrading and Renovation Project of Civil Service Hospital, and Mutually agreed Post-Disaster Reconstruction Projects.[132] In 2019 the Chinese President Xi Jinping visited Nepal during which the two countries signed a number of MoUs.

India-Bhutan relationships

For the Bhutanese, India is *gyagar* (the holy land) because Buddhism which is being practiced by them was born in India. Buddhism was popularized in Bhutan by an Indian monk named Padmasambhava.[133] In Bhutan he is popularly known as Guru Rinpoche.[134]

   In medieval India, especially after the spread of Mughal rule in the Indian State of Bengal, confrontations to control the fertile plains of Bengal between the Mughal and Bhutan used to occur. During the East India Company's rule in India, there were frequent skirmishes between the Bhutanese and the Company's army over *duars* (fertile land in Rivers Brahmaputra and Ganga region in Assam and Bengal).[135] After the defeat of Bhutanese in clash with the East India Company's troops in 1773, the two clashing parties signed a treaty of peace and commerce in 1774.[136] However, peace was finally established only after the signing of the Treaty of Sinchula between Bhutan and British India in November 1865. Under this treaty Bhutan ceded a part of *duars* to British India. In 1910, to address China's charges in Bhutanese territory, the Treaty of Punakha was signed between Bhutan and British India. Under this treaty, the British agreed not to interfere in the internal affairs of Bhutan to which the Bhutanese agreed "to be guided by the advice of India in regard to its foreign relations".[137]

After independence in 1947, on 8 August 1949 India and Bhutan signed the Treaty of Perpetual Peace and Friendship. In 1968 a diplomatic relationship between India and Bhutan was established with an appointment of a resident representative of India in Thimphu. Prior to it a political officer in Sikkim was in charge of India's Bhutan relations.[138] In the late 1960s, 70s, and 80s there were moments when it seemed that Bhutan was moving away from India. For example, in 1968, Bhutan barred unauthorised foreigners, including Indians, from entering into its territory. A year later in 1969, Bhutan also started its own currency.[139] In 1970, after setting up its ministry for foreign affairs, Bhutan tried to have an independent foreign policy.[140] In 1980s Bhutan also reduced India's annual assistance to Bhutan by 43% and started inviting other international donors.[141] However, the bilateral relationships between India and Bhutan was back on track after the visit of the Indian Prime Minister Rajiv Gandhi in 1985 and September 1988 and then by the Indian President R. Venkatraman in October 1988.[142]

In March 2007, the Indo-Bhutan friendship treaty was updated at Thimphu. The updated treaty addresses economic and security related interests of both signatories. More important, the 2007 updates have removed some of the provisions continued for long under the 1949 India-Bhutan friendship treaty. Article 2 of the treaty states

> The Government of India undertakes to exercise no interference in the internal administration of Bhutan. On its part the Government of Bhutan agrees to be guided by the advice of the Government of India in regard to its external relations.[143]

Under Article 6 of the 1949 Treaty

> The Government of India agrees that the Government of Bhutan shall be free to import with the assistance and approval of the Government of India, from or through India into Bhutan, whatever arms, ammunition, machinery, warlike material or stores may be required or desired for the strength and welfare of Bhutan, and that this arrangement shall hold good for all time as long as the Government of India is satisfied that the intentions of the Government of Bhutan are friendly and that there is no danger to India from such importations. The Government of Bhutan, on the other hand, agrees that there shall be no export of such arms, ammunition, etc., across the frontier of Bhutan either by the Government of Bhutan or by private individuals.[144]

Article 2 of the 2007 updated treaty states that

> In keeping with the abiding ties of close friendship and cooperation between Bhutan and India, the Government of the Kingdom of Bhutan and the Government of the Republic of India shall cooperate closely with each other on issues relating to their national interests. Neither Government shall allow the use of its territory for activities harmful to the national security and interest of the other.[145]

On security front, Article 4 of the Treaty states that

> The Government of India agrees that the Government of Bhutan shall be free to import, from or through India into Bhutan, whatever arms, ammunition, machinery, warlike material or stores as may be required or desired for the strength and welfare of Bhutan, and that this arrangement shall hold good for all time as long as the Government of India is satisfied that the intentions of the Government of Bhutan are friendly and that there is no danger to India from such importations. The Government of Bhutan agrees that there shall be no export of such arms, ammunition and materials outside Bhutan either by the Government of Bhutan or by private individuals.[146]

Article 2 of the updated treaty relieve Bhutan from guidance from India. Then Article 3 and 8 of the treaty calls for having free trade and commerce, and consolidation of economic relationships between the two countries.[147]

After the update of the India-Bhutan friendship treaty, in 2008, then Indian Prime Minister Dr Manmohan Singh visited Bhutan. At that time India agreed to let Bhutan have 16 entry and exit points to facilitate trade and commerce with other countries except China. In 2014, soon after becoming Prime Minister of India, Narendra Modi paid a visit to Bhutan. Modi inaugurated the India-funded building of the Supreme Court of Bhutan.

In 2016 India and Bhutan signed an agreement on trade, commerce, and transit between the two countries. Under this transit and trade agreement it has been agreed that there shall be free trade and commerce between India and Bhutan. Notwithstanding, Bhutan may impose such non-tariff restrictions on the entry into Bhutan of certain goods of Indian origin as may be necessary for the protection of industries in Bhutan. Such restrictions, however, will not be stricter than those applied to goods of third country origin. However, the two Governments may impose such non-tariff restrictions on entry into their respective territories of goods of third country origin as may be necessary.[148] In 2018 Bhutanese consulate was opened at Guwahati.

Although it seems that India and Bhutan share cordial relationships, there are certain issues over which differences exist between the two countries. For example, in April 2017 Bhutan pulled out from India's initiative for having a Bangladesh, Bhutan, India, and Nepal Motor Vehicle Agreement. This was done by the Bhutanese government with an eye on the then upcoming elections in the country in 2018.[149]

Like other South Asian countries, China has not been able to establish itself firmly in Bhutan, though it has been trying to do so from years. The official contact between Bhutan and China began in 1970s.[150] In 1974, along with few other countries, China was also invited to the coronation of King Jigme Singye Wangchuck. The Chinese delegation was led by Ma Muming, Chargé d'Affaires of the Chinese embassy in New Delhi.[151] In 1979, as a result of Chinese intrusions into Bhutanese territory, the Royal Government of Bhutan thought about direct negotiations with China on the boundary issue.[152] As a part of the process, in 1983, Chinese State Councillor and Foreign Minister Wu Xueqian and Bhutanese Foreign Minister Dawa Tsering met in New York and held consultations on developing bilateral relations.[153] The first round of boundary talks was held in Beijing in April 1984. From 1994, Chinese ambassadors to India have been paying working visits to Bhutan. Since 1995, Bhutan has been "supporting (the) one-China policy".[154] In 1998 Bhutan and China signed an agreement on the Maintenance of Peace and Tranquillity Along the Sino-Bhutanese Border Areas. The two sides accepted that they have reached an agreement and agreed to work in accordance with the five principles of mutual respect for each other's sovereignty and territorial integrity, mutual non-aggression, mutual non-interference in each other's internal affairs, and peaceful co-existence and for the purpose of maintaining peace and tranquillity along the Sino-Bhutanese border.[155] In June 2000, the Bhutanese Ambassador to India visited China. These visits have opened up a new channel of contacts other than the boundary talks. In 2015, the 23rd round of talks were held in Thimphu and, in 2016, technical expert groups from the two countries met in Bhutan.[156] Despite agreement and ongoing talks, in 2017 a military stand-off took place between India and China when China laid a territorial claim over Doklam/Doko-La (or Donglong) in Bhutan. The territory in focus was a plateau of approximately 89 square kilometres, which is at the tri-junction of India, China, and Bhutan. It is close to India's "Chicken's Neck", the Siliguri Corridor. The border remained tense for about 73 days, following which India and China agreed to disengage their personnel from that site on 28 August 2017.

The Chinese government claimed that the land is located on its side of the border as per the 1890 Convention between Great Britain and China relating to Sikkim and Tibet and it was, therefore, free to construct a road near the site. However, on 29 June 2017, in a press release, the Bhutanese government stated:

> Boundary talks are ongoing between Bhutan and China and we have written agreements of 1988 and 1998 stating that the two sides agree

to maintain peace and tranquillity in their border areas pending a final settlement on the boundary question, and to maintain status quo on the boundary as before March 1959. The agreements also state that the two sides will refrain from taking unilateral action, or use of force, to change the status quo of the boundary. Bhutan has conveyed to the Chinese side, both on the ground and through the diplomatic channel, that the construction of the road inside Bhutanese territory is a direct violation of the agreements and affects the process of demarcating the boundary between our two countries. Bhutan hopes that the status quo in the Doklam area will be maintained as before 16 June 2017.[157]

Post-military stand-off at Doklam, Assistant Foreign Minister of China, Kong Xuanyou, visited Bhutan from 22 to 24 July 2018. He was accompanied by Luo Zhaohui, the Chinese Ambassador to India. In Bhutan, the Assistant Foreign Minister met the fifth king, Jigme Khesar Namgyel Wangchuck, the fourth king, Jigme Singye Wangchuck, and Tobgay.[158] During the meetings, the Chinese Assistant Foreign Minister stated,

> The Chinese side is willing to work with the Bhutanese side to maintain high-level contacts, expand practical cooperation, and strengthen multilateral communication and coordination, so as to achieve common development on the basis of mutual respect, mutual benefit and win-win results. The Chinese side welcomes Bhutan to actively participate in the Belt and Road Initiative and share China's development dividend. Both sides should continue to promote the boundary negotiations, abide by the already-reached principles and consensus, and jointly maintain peace and tranquillity in boarder areas so as to create positive conditions for the final settlement of the boundary issue.[159]

On its part,

> The Bhutanese side expressed that, although China and Bhutan have not formally established diplomatic relations, the two countries have enjoyed a traditional friendship which could be regarded as a model between big countries and small countries. The Bhutanese side admires the development achievements of China and welcomes the positive outcomes of the Belt and Road Initiative proposed by President Xi Jinping . . . Bhutan firmly adheres to the one-China policy and is committed to deepening exchanges and cooperation with China, and stands ready to maintain communication with the Chinese side on bilateral relations and the boundary issue.[160]

About the Chinese Assistant Foreign Minister's visit to Bhutan, it seems that Thimpu kept New Delhi "in the loop".[161]

Conclusion

South Asia as a region includes those countries who are ethnically, culturally, and geographically contiguous to India and each other. Although the definitions of the region, as seen, keep on changing some similarities other than the sharing of a border are utmost. There are scholars and political commentators who made an argument for including China while talking about South Asian resource politics or security, they may have their argument, and so have many in India's neighbourhood, however, contiguity is one of the many factors that defines a region.

Having said that, one cannot deny the fact that China is a part of hydrological South Asia. Most of the major rivers of the region originates from Tibet, in China. Hence, any basin-related solution to South Asian waters cannot ignore China. Second, as an external actor to South Asia, China is engaged in the construction, operation, and maintenance of a number of water projects in India's riparian neighbours[162] except Bhutan. Such engagements make China to influence the water policy related decisions, particularly in Pakistan, and to some extent in Nepal and Bangladesh.

Finally, the bilateral relationships between the riparian neighbours may not have a direct impact on the transboundary rivers water issues but they do influence their water-related policies and behaviours. Often good bilateral relationships help the riparian neighbours to sign water agreements or treaties. It also helps them to amicably resolve their differences over water distribution and hydroelectricity projects on the shared rivers. In South Asia, India, and Bangladesh signed a Ganges Water Sharing treaty in 1996, mainly because of their improved bilateral relationships, when Hasina came into power. Then India has managed to acquire contracts to build hydroelectric power projects in Nepal and Bhutan because of its relationships with the two respective countries. Therefore, it is of utmost importance to know and understand bilateral relationships before discussing India's transboundary rivers waters disputes with its South Asian riparian neighbours. This chapter has attempted to briefly examine and understand the nature of bilateral relations India has with its South Asian neighbours.

Notes

1 Hagrety, Devin T. and Herbert G. Hagerty. (2007). " Introduction" & "Reconstitution and Reconstruction of Afghanistan". In Hagrety, Devin T. (ed) *South Asia in World Politics*. Lanham, Boulder, New York, Toronto and Oxford: Rowman and Littlefield Publishers Inc, pp. 4 and 113–133.
2 See, Noorani, A.G (2014) *The Destruction of Hyderabad*. London: C.Hurst.
3 Scott, David. (2013). "India's 'Extended Neighbourhood' Concept: Power Projection for a Rising Power". In Bajpai, Kanti P. and Harsh Pant (eds) *India's Foreign Policy: A Reader*. New Delhi: Oxford University Press, pp. 321–348.
4 Chacko, Priya. (2012). *Indian Foreign Policy: The Politics of Postcolonial Identity from 1947 to 2004*. London and New York: Routledge.

5 Buzan, Barry and Ole Waever. (2003). *Regions and Powers: The Structure of International Security*. London: Cambridge University Press, p. 44.
6 Chari, P.R. and V. Raghvan. (2010). *Comparative Security Dynamics in North East Asia and South Asia*. New Delhi: Pentagon.
7 Butalia, Urvashi. (2014). *The Other Side of Silence: Voices from the Partition of India*. New Delhi: Penguin Books; Bhattacharya, Sabyasachi. (2014). *The Defining Moments in Bengal, 1920–1947*. New Delhi: Oxford University Press; Bhalla, Alok. (2006). *Partition Dialogues: Memories of a Lost Home*. New Delhi: Oxford University Press.
8 Butalia, U. (2007). *The Other Side of Silences: Voices From the Partition of India*. New Delhi: Penguin Books.
9 Ranjan, Amit. (2016) "Disputed Waters: India, Pakistan and the Transboundary Rivers". *Studies in Indian Politics* (Sage and Centre for the Study of Developing Societies, New Delhi), Vol. 4, No. 2, pp. 191–205.
10 Hussian, Z. (2010). *Frontline Pakistan: The Path to Catastrophe and the Killing of Benazir Bhutto*. New Delhi: Viva Books; Kiessling, Hein G. (2016). *Faith, Unity, Discipline: The ISI of Pakistan*. New Delhi: Harper Collins.
11 It comprises Assam, Meghalaya, Manipur, Sikkim, Nagaland, Arunachal Pradesh and Mizoram. Among these States except Sikkim which became a part of India in 1975 all other States have witnessed sub-nationalist and secessionist movements led by local militia groups. At present, most of the such movements have died down.
12 Hauzel, H. (2005). Arms, Drug Smuggling and Cross Border Terrorist Activities. In Shoban, F. (ed) *Dynamics of Bangladesh-India Relations: Dialogue of Young Journalists Across the Border*. Dhaka: Bangladesh Enterprise Institute, The University Press, pp. 39–54.
13 Jacob, Jayant. (2017, 2 September). "Pakistan Supporting Sikh Militants, Say Fresh Intelligence Inputs". *Hindustan Times*. www.hindustantimes.com/india-news/pakistan-supporting-sikh-militants-say-fresh-intelligence-inputs/story-KJTEQPdHMj9FBawDC2gOrN.html. Accessed on 19 March 2018.
14 Dulat, A.S. and A. Sinha. (2015). *Kashmir: Vajpayee Years*. New Delhi: Harper & Collins.
15 "Who Is Kulbhushan Jadhav?", *Dawn*, 2017, 18 May. www.dawn.com/news/1326117. Accessed on 22 May 2017.
16 "Who Is Kulbhushan Jadhav?" *The Hindu*, 2017, 10 April. www.thehindu.com/news/ national/who-is-kulbhushan-jadhav-the-hindu-explains/article17907888.ece. Accessed on 22 May 2017.
17 Ministry of External Affairs, Government of India. "Annual Report 2017–18". www.mea.gov.in/Uploads/PublicationDocs/29788_MEA-AR-2017-18-03-02-2018.pdf, p. Vii.
18 "Pakistani PM Imran Khan Brands India 'Arrogant' After It Cancels Talks". *The Strait Times*, 2018, 24 September. www.straitstimes.com/asia/pakistani-pm-khan-brands-india-arrogant-after-it-cancels-talks. Accessed on 24 September 2018.
19 "Timeline of dialogue process between India, Pakistan", *Business Standard* 2015, 22 August. https://www.business-standard.com/article/news-ians/timeline-of-dialogue-process-between-india-pakistan-115082200662_1.html. Accessed on 28 August 2018.
20 Ibid.
21 Ibid.
22 Ibid.
23 "Pakistan, India peace talks 'suspended'". *BBC News* 2016, 8 April. https://www.bbc.com/news/world-asia-india-35994599. Accessed on 18 June 2017.

24 " 'Terror and talks can't go together': Sushma Swaraj clarifies on resumption of dialogue with Pakistan" *Financial Express*, 2018, 28 November. https://www.financialexpress.com/india-news/terror-and-talks-cant-go-together-sushma-swaraj-clarifies-on-resumption-of-dialogue-with-pakistan/1396571/. Accessed on 20 July 2019.
25 "Pak Parties Slam Imran Khan For "Too Much Keenness" On Talks With India" *NDTV* 2018, 23 September. https://www.ndtv.com/world-news/paki stan-opposition-parties-slam-imran-khan-for-too-much-keenness-on-talks-with-india-1920851. Accessed on 19 June 2019.
26 Roy, Shubhajit (2019, 1 September) "Open to conditional talks with India: Pakistan foreign minister" *The Indian Express*. https://indianexpress.com/arti cle/india/open-to-conditional-talks-with-india-pakistan-foreign-minister-shah-mahmood-qureshi-5955165/. Accessed on 1 September 2019.
27 Sahni, Varun. (2001). "Technology and Conflict Resolution: The Siachen Conflict". In Sahadevan, P. (ed) *Conflict and Peacemaking in South Asia*. New Delhi: Lancers Book, pp. 236–271.
28 Ibid.
29 Ibid.
30 Ibid.
31 Raghavan, V.R. (2002). *Siachen: Conflict Without End*. New Delhi: Viking, p. 129.
32 "Siachen Was Almost a Done Deal". *The Hindu*, 2012, 9 June. www.thehindu.com/news/national/siachen-was-almost-a-done-deal-in-1992/article3509787.ece. Accessed on 12 January 2018.
33 Ibid.
34 "Indian Soldiers Killed After Avalanche Hits Siachen Glacier". (2010, 19 November). *Al Jazeera*. https://www.aljazeera.com/news/2019/11/indian-sol diers-killed-avalanche-hits-siachen-glacier-191119062742949.html. Accessed on 27 November 2019.
35 Shah, Sikander Ahmed. (2015, 23 February). "Without a Peddle". *Dawn*. https://www.dawn.com/news/1165267. Accessed on 18 September 2018.
36 Ibid.
37 Ibid.
38 Dabas, Maninder. (2016, 16 August). "Everything You Need to Know About the Dispute Over Sir Creek Between India and Pakistan". *India Today*. https://www.indiatimes.com/news/everything-you-need-to-know-about-the-dispute-over-sir-creek-between-india-and-pakistan-260071.html. Accessed on 18 September 2017.
39 Ibid.
40 According to thalweg principles, the border between two states separated by a watercourse or flowing body of water as lying along the thalweg (meaning valley way), which is the line of greatest depth of the channel or watercourse.
41 Gupta, A.K. and P. Sahadevan. (2001). *Conflict and Peacemaking in South Asia*. New Delhi: Lancers Book, pp. 272–295.
42 Ibid.
43 Dabas, Maninder. (2016, 16 August). "Everything You Need to Know About the Dispute Over Sir Creek Between India and Pakistan". *India Today*. https://www.indiatimes.com/news/everything-you-need-to-know-about-the-dispute-over-sir-creek-between-india-and-pakistan-260071.html. Accessed on 18 September 2017.
44 Many of the Mujahideen from Afghan theatre shifted their activity to the Indian-administered Kashmir after the withdrawal of the Soviet Union from Afghanistan. Militant fundamentalist organisations in Pakistan such as the Lashkar-e-Taiba

and Jaish-e-Muhammad began to recruit and train volunteers for the Kashmir jihad (Cited in Ahmed, Ishtiaq. (2009). "Spectre of Islamic Fundamentalism". In Jetly, Rajshree (ed) *Pakistan in Regional and Global Politics*. London, New York and New Delhi: Taylor & Francis Group, pp. 150–180; Hussian, Zahid. (2010). *Frontline Pakistan: The Path to Catastrophe and the Killing of Benazir Bhutto*. New Delhi: Viva Books, p. 24).

45 Roche, Elizabeth. (2019, 24 September). "Khan Admits Pakistan Army, ISI Trained Militant Groups in Kabul". *Live Mint*. https://www.livemint.com/news/world/imran-khan-admits-pakistan-army-isi-trained-al-qaeda-1569312571984.html. Accessed on 19 October 2019.

46 Nawaz, Shuja. (2019). *The Battle for Pakistan: The Bitter US Friendship and a Tough Neighbourhood*. Maryland: Rowman & Littlefield.

47 Hudobhoy, P. (2011, 25 June–8 July). "Pakistan Army Divided It Stands". *Economic & Political Weekly*, Vol. XLVI, No. 26–27, pp. 67–71.

48 "Pakistan Taliban: Peshawar School Attack Leaves 141 Dead". *BBC*, 2014, 16 December. https://www.bbc.com/news/world-asia-30491435. Accessed on 19 December 2017.

49 Dulat, A.S. and A. Durrani. (2011, July 14). "India-Pakistan Intelligence Cooperation". *The Hindu*.

50 Ministry of External Affairs, Government of India " India-Pakistan Joint Anti-Terrorism Mechanism to hold its first meeting on 06 March, 2007" https://mea.gov.in/press-releases.htm?dtl/2296/IndiaPakistan+Joint+AntiTerrorism+Mechanism+to+hold+its+first+meeting+on+06+March+2007. Accessed on 14 March 2017.

51 "Pakistan Says It Has Asked India To Provide More Evidence In 26/11 Trial" *NDTV* 2016, 30 June. https://www.ndtv.com/india-news/pakistan-says-it-has-asked-india-to-provide-more-evidence-in-26-11-trial-1426552. Accessed on 31 July 2017.

52 "Pak Media Taunts Indian Security Agencies". *Hindustan Times*, 2008, 3 December. https://www.hindustantimes.com/india/pak-media-taunts-indian-security-agencies/story-yi8F5eQYXva8BPDNSFOF5I.html. Accessed on 25 July 2019.

53 Rashid, A. (2011). "The Afghan Conundrum". In Lodhi, Maleeha (ed) *Pakistan: Beyond the Crisis State*. New Delhi: Rupa, pp. 305–317.

54 Jones, Bennet Owen. (2015, 25 June). "Pakistan's MQM 'Received Indian Funding'". *BBC News*. https://www.bbc.com/news/world-asia-33148880. Accessed on 18 March 2018.

55 Bhaumik, Subir. (2015, 2 July). "Raking Up the MQM-RAW Link for Brownie Points". *The Hindu*. https://www.thehindu.com/opinion/op-ed/raking-up-the-mqmraw-link-for-brownie-points/article7375459.ece. Accessed on 12 March 2018.

56 Kamal, Sayid. (2015, 18 February). "Blow the Pak Boat Off, We Do Not Want to Serve Them Biryani: COAS Guard DIG". *The Indian Express*. https://indianexpress.com/article/india/india-others/i-told-at-night-blow-the-pak-boat-off-we-dont-want-to-serve-them-biryani-coast-guard-dig/.Accessed on 19 February 2018.

57 Singh, Uma. (2001). "Kashmir: The 'Core Issue' Between India and Pakistan". In Sahadevan, P. (ed) *Conflicts and Peacemaking in South Asia*. New Delhi: Lancers Book, p. 214.

58 Bilkenberg, L. (1998). *India-Pakistan: The History of Unsolved Conflicts (vol. II): Analyses of Some Structural Factors*. Campusvej: Odense University Press, 1998.

59 Khursheed Mohammad Kasuri during his interaction with Times of India and Jang group's Aman ki Asha programme on 23 April 2010. Statement published on 24th April 2010.

60 Bose, Sumantra. (2005). *Kashmir: Roots of Conflict, Paths to Peace*. Harvard: Harvard University Press, pp. 48–50.
61 Donthi,Praveen (2016, 23 March) " How Mufti Mohammad Sayeed Shaped the 1987 Elections in Kashmir" *The Carvan*. https://caravanmagazine.in/van tage/mufti-mohammad-sayeed-shaped-1987-kashmir-elections. Accessed on 18 May 2017.
62 Bose, Sumantra. (2005). *Kashmir: Roots of Conflict, Paths to Peace*. Harvard: Harvard University Press, pp. 48–50.
63 Masood, Basharaat and Nirupama Subramanian. (2017, 14 April). "Two Percent Turnout in Sri Nagar Repoll, Zero in 20 Booths: 'I Voted But Govt Must Understand Anger'". *The Indian Express*. http://indianexpress.com/article/india/2-turnout-in-srinagar-repoll-zero-in-20-booths-i-voted-but-govt-must-understand-the-anger-4612411/. Accessed on 17 April 2017.
64 Noorani, A.G. (2019, 13 August). "Murder of Insaniyat, and of India's Solemn Commitment to Kashmir". *The Wire*. https://thewire.in/law/murder-of-insaniyat-and-of-indias-solemn-commitment-to-kashmir. Accessed on 13 August 2019.
65 Schofield, V. (2010). *Kashmir in Conflict: India, Pakistan and the Unending War* (1st South Asian ed.). London and New York: I. B. Tauris.
66 "Revoking 370 is 'murder of Democracy' say left parties" *Businessline* 5 August 2019. https://www.thehindubusinessline.com/news/national/revoking-article-370-is-amurder-of-democracy-say-left-parties/article28821850.ece. Accessed on 12 August 2019.
67 Sanaullah Khan, Naveed Siddiqui and Tahir Sherani. (2019, 7 August). "Pakistan Suspends Trade With India, Asks Indian Envoy to Leave". *Dawn*. https://www.dawn.com/news/1498609. Accessed on 1 August 2019.
68 "Attari Integrated Check Post to Open for Trade on Friday". *The Hindu*, 2012, 12 April. www.thehindu.com/news/national/attari-integrated-check-post-to-open-for-trade-on-friday/article3304431.ece. Accessed on 18 July 2019.
69 Ikram Junadi. (2020, 6 September). "Ban on Import of Indian Medicines Lifted". *Dawn*. https://www.dawn.com/news/1503357. Accessed on 12 September 2019.
70 Iqbal, Shahid. (2020, 23 January). "Indo-Pak Trade Plunges in Six Months". *Dawn*. www.dawn.com/news/1530016/indo-pak-trade-plunges-in-six-months. Accessed on 24 January 2020.
71 After the assassination of Sheikh Mujibur Rahman, the Bangladesh military took charge of Bangladesh. For a brief period, they ruled from behind the curtains.
72 Ministry of External Affairs, Government of India. "India and Bangladesh: Land Boundary Agreement". www.mea.gov.in/Uploads/. . ./24529_LBA_MEA_Book let_final.pdf. Accessed on 17 April 2017.
73 Tea plantation in Assam was introduced by Scottish, Robert Bruce. He started company which expanded trade of Assam tea to the other parts of the world. When Bruce landed in Assam, he discovered tea plants 'growing wild in the upper Brahmaputra valley'. See "History of Indian Tea". *Indian Tea Association*. www.indiatea.org/history_of_indian_tea. Accessed on 3 August 2018.
74 Gait, Edward. (1926). *A History of Assam*. Calcutta and Shimla: Thacker, Spink & CO.
75 Ibid.
76 Datta, A. (2013). *Refugees and Borders in South Asia: The Great Exodus of 1971* London: Routledge.
77 For details on references on IMDT, see www.india-eu-migration.eu/media/legalm odule/Illegal%20 Migrants%20Act%201983.pdf. Accessed on 4 August 2018.
78 Basu, Ranjan. (2018, 25 May). "How Sheikh Hasina Handled Jyoti Basu Over Bangladeshi Intrusion Issue". *Dhaka Tribune*. www.dhakatribune.com/

bangladesh/foreign-affairs/2018/05/25/how-sheikh-hasina-handled-jyoti-basu-over-bangladeshi-intrusion-issue. Accessed on 29 June 2019.

79 Ibid.
80 "Assam Publishes First Draft of NRC with 1.9 Crore Names". *The Times of India*, 2018, 1 January. https://timesofindia.indiatimes.com/city/guwahati/assam-publ ishes-first-draft-of-nrc-with-1-9-crore-names/articlesh ow/62322518.cms. Accessed on 19 May 2018.
81 Agarwala, Tora. (2018, 30 July). "Assam Citizenship List: Names Missing in NRC Final Draft, 40 Lakh Ask What Next". *The Indian Express*. https:// indianexpress.com/article/north-east-india/assam/assam-citizenship-list-names-missing-in-nrc-final-draft-40-lakh-ask-what-next-5283663/. Accessed on 30 July 2018.
82 "No Worries Over Assam NRC Draft". *The Daily Star*, 2018, 2 August. www. thedailystar.net/back page/nothing-worry-about-over-assam-nrc-draft-1614724. Accessed on 4 August 2018.
83 "Quietly, Delhi Kept Dhaka in NRC Loop: No Deportation Talk". *The Indian Express*, 2018, 4 August. https://i ndianexpress.com/article/india/quietly-delhi-kept-dhaka-in-nrc-loop-no-deportation-talk-5291048/. Accessed on 4 August 2018.
84 Ibid.
85 Ibid.
86 "BJP Plays Politics on 'Ousting Infiltrators', Deportation Data Tells Different Story". *The Wire*, 2018, 3 August. www.thewire.in/politics/amit-shah-bjp-nrc-assam-upa. Accessed on 4 August 2018.
87 Government of Assam, Office of the State Coordinator of National Registration (NRC) Assam. "Writ Petition (Civil) NO. 562 of 2012, Writ Petition (Civil) NO. 274 of 2009 & Writ Petition NO. 876 of 2014". Government of Assam, Office of the State Coordinator of National Registration (NRC) Assam. www.nrcas sam.nic.in/images/pdf/01.pdf. Accessed on 28 May 2018, p. 57.
88 Ministry of Law and Justice, Government of India. "Citizenship Amendment Act, 2019". http://egazette.nic.in/WriteReadData/2019/214646.pdf. Accessed on 12 January 2020.
89 Protestors told the author that it is discriminatory and linked with NRC.
90 "No Illegal Immigrant Can Stay in Assam or Enter Other States: Amit Shah". NDTV. 2019, 9 September. https://www.youtube.com/watch?v=6-9RI2LvY9k. Accessed on 12 October 2019.
91 Roy, Shubajit. (2020, 7 March). "Ahead of Modi's Dhaka Trip, Protests Held Against CAA". *The Indian Express*. https://indianexpress.com/article/world/pm-modi-dhaka-trip-citizenship-amendment-act-caa-protests-6303034/. Accessed on 7 March 2020.
92 Kabir, Mahfuz. (2016, 10 October). "Expanding the Bangladesh-China Trade Frontier". *The Daily Star*. www.thedailystar.net/op-ed/politics/expanding-the-bangladesh-china-trade-frontier-1296583. Accessed on 15 October 2016.
93 "BD, China Set to Sign Over 25 MoUs, Deals". *The Financial Express* 2016, 13 October. www.thefinancialexpress-bd.com/2016/10/13/49214/BD,-China-set-to-sign-over-25-MoUs,-deals. Accessed on 14 October 2016.
94 "China- Bangladesh Cooperation Will Bear Golden Fruits". *The Daily Star*, 2016, 14 October. www.thedailystar.net/frontpage/china-bangladesh-coopera tion-will-bear-golden-fruits-1298536. Accessed on 15 October 2016.
95 Sakhuja, Vijay. (2009, 23 July). "China-Bangladesh Relations and Potential for Regional Tensions". *China Brief*, Vol. 9, No. 15. www.jamestown.org/single/?tx_ ttnews%5Btt_news%5D=35310&no_cache=1#.V2E4A9L5jIU. Accessed on 15 June 2016.

96 Ibid.
97 "China, Bangladesh Lift Ties to Strategic Partnership of Cooperation". (2016, 15 October). *Global Times*. www.globaltimes.cn/content/1011533.shtml. Accessed on 16 October 2016. Also Ministry of Foreign Affairs, Government of Bangladesh. "Joint Statement". http://mofa.gov.bd/media/joint-statement-people%E2%80%99s-republic-china-and-people%E2%80%99s-republic-bangladesh-establishing. Accessed on 18 October 2016.
98 Majumder, Shantanu. (2016). "Secularism and Anti-Secularism". In Ali, Riaz and Mohammad Sajjadur Rahman (eds) *Routledge Handbook of Contemporary Bangladesh*. Oxon and New York: Routledge, pp. 40–51.
99 Ahmed Chowdhury, Iftekhar. (2016, 4 July). "Terror in Dhaka: Fundamentalism Spreads Its Deadly Wings". *ISAS Brief No 439*. www.isas.nus.edu.sg/wp-content/uploads/media/isas_papers/ISAS%20Brief% 20No.%20439%20-%20 Terror%20in%20Dhaka%20-%20Fundamentalism%20Spreads%20its%20 Deadly% 20Wings.pdf. Accessed on 7 May 2018.
100 Ranjan, Amit and Roshini Kapoor. (forthcoming). "Militancy in Bangladesh". In S. Narayan and Srreradha Dutta (eds) *Assessing Transformation of Bangladesh* (Tentative). Hyderabad: Orient Longman.
101 Roy, Shubhajit. (2016, 3 July). "Bangladesh Café Attack: Indian Among 20 Killed in Dhaka's Night of Terror". *The Indian Express*. http://indianexpress.com/article/world/world-news/bangladesh-cafe-attack-indian-among-20-killed-in-dhakas-night-of-terror-2890521/. Accessed on 5 July 2016.
102 Nair, Rupam Jain and Andrew MacAskil. (2014, 28 October). "India Uncovers Suspected Plot to Assassinate Bangladeshi PM-Security Officials". *Reuters*. https://in.reuters.com/article/india-bangladesh-plot-hasina/india-uncovers-suspected-plot-to-assassinate-bangladeshi-pm-security-officials-idINKBN0I H1HN20141028. Accessed on 18 November 2017.
103 Gupta, Shishir. (2019, 31 May). "Bangladesh Terror Group Setting Up Bases Near Border, Ties Up With Pak's LeT". *Hindustan Times*. https://www.hindustantimes.com/india-news/jmb-joins-hands-with-let-to-expand-activities-in-india/story-Wf0FqoCAfwZNW0ntDTYB9J.html. Accessed on 25 May 2019.
104 "Together Against All Terrorism". (2016, 3 May). *The Daily Star*. www.thedailystar.net/frontpage/together-against-all-terrorism-1222996.
105 Muni, S.D. (2015). "India's Nepal Policy". In Malone, David M., C. Raja Mohan and R. Srinath (eds) *The Oxford Handbook of Indian Foreign Policy*. New Delhi: Oxford University Press, pp. 398–411.
106 "Sugauli Treaty of 1816: Full Text". http://nepaldevelopment.pbworks.com/w/page/34197552/Sughauli%20Treaty%20of%201815%3A%20Full%20Text#MemorandumfortheapprovalandacceptanceoftheRajaofNipal. Accessed on 9 November 2019.
107 Bhasin, A.S. (2005). *Nepal-India, Nepal-China Relations Documents 1947-June 2005 Vol-I*. New Delhi: Geeta Press, p. 94.Ministry of External Affairs, Government of India 'Treaty of Peace and Friendship', 31 July 1950. https://mea.gov.in/bilateral-documents.htm?dtl/6295/Treaty+of+Peace+and+Friendship. Accessed on 12 May 2018.
108 Ibid.
109 A relationship based on marriage, family and food. As the people from terai and Indian state of Bihar shares common caste identity they enter into marital relationship. Even the erstwhile monarchs, prince and princess from India and Nepal had been related on the basis of marriage.
110 Malhotra, Jyoti. (2015, 4 November). "Troubled Transition". *India Today Web Report*. http://indiatoday.intoday.in/story/troubled-transition/1/516255. html. Accessed on 6 November 2015.

111 Whelpton, J. (2005). *A History of Nepal*. Cambridge: Cambridge University Press, p. 98.

112 Bhasin, Avtar Singh. (2005). *Nepal-India, Nepal China Relations: Documents 1947–June 2005 Volume 1*. New Delhi: Geetika Publications.

113 Ibid.

114 Ibid., p 523.

115 Cown, Sam. (2015, 14 December). "The Indian Checkposts, Lipu Lekh, and Kalapani". *The Record*. https://www.recordnepal.com/wire/indian-checkposts-lipu-lekh-and-kalapani/. Accessed on 14 November 2019.

116 Muni, S.D. (2015). "India's Nepal Policy". In Malone, David M., C. Raja Mohan and R. Srinath (eds) *The Oxford Handbook of Indian Foreign Policy*. New Delhi: Oxford University Press, pp. 398–411.

117 Ibid.

118 Bhattarai, Tek Narayan (2015, 5 October) "Remembering the 1989 Blockade" *Nepali Times*. http://archive.nepalitimes.com/article/from-nepali-press/Remembering-the-1989-blockade,2651. Accessed on 12 March 2019.

119 Bhasin, A.S. (2005). *Nepal-India, Nepal – China Relations Documents 1947-June 2005 Volume II*. New Delhi: Geeta Press, p. 797.

120 It included Nepali Congress, Nepal Congress (Democratic), Communist Party of Nepal (United Marxist-Leninist), Nepal Workers and Peasants Party, Nepal Goodwill Party (Anandi Party), United Left Front and People Front.

121 Roche, Elizabeth. (2016, 9 August). "Nepal's Prachanda and India: A Love-Hate Relationship". *Livemint*. https://www.livemint.com. Accessed on 17 April 2018.

122 Bhattacharya, Kallol. (2015, November). "Nepal PM Draws Ire for Anti-India Speech". *The Hindu*. https://www.thehindu.com/news/international/nepal-pm-oli-draws-ire-for-antiindia-speech/article7884613.ece. Accessed on 18 December 2018.

123 Pattison, Pete. (2015, 6 November). "Nepal Warns of Humanitarian Crisis as India Border Blockade Continues". https://www.theguardian.com/global-development/2015/nov/06/nepal-government-humanitarian-crisis-india-border-blockade-deputy-prime-minister-kamal-thapa. Accessed on 12 January 2018.

124 Jain, Bharti. (2019, 2 November). "Govt Releases New Political Map of India Showing UTs of J&K, Laddakh". *The Times of India*. https://timesofindia.indiatimes.com/india/govt-releases-new-political-map-of-india-showing-uts-of-jk-ladakh/articleshow/71867468.cms. Accessed on 20 November 2019.

125 "India's Updated Political Map Stirs Controversy in Nepal". *Al Jazeera*, 2019, 8 November. www.aljazeera.com/news/2019/11/india-updated-political-map-stirs-controversy-nepal-191108130802391.html. Accessed on 9 November 2019.

126 "New Delhi Responds to Uproar in Nepal Over New Political Map, Says It Is 'Accurate' ". *The Kathmandu Post*, 2019, 7 November. https://kathmandupost.com/politics/2019/11/07/new-delhi-responds-to-uproar-in-nepal-over-new-political-map-says-it-is-accurate. Accessed on 9 November 2019.

127 Ibid.

128 Ministry of Foreign Affairs, Government of Nepal. "Press Release". https://mofa.gov.np/%e0%a4%aa%e0%a5%8d%e0%a4%b0%e0%a5%87%e0%a4%b8-%e0%a4%b5%e0%a4%bf%e0%a4%9c%e0%a5%8d%e0%a4%9e%e0%a4%aa%e0%a5%8d%e0%a4%a4%e0%a4%bf-%e0%a4%a8%e0%a5%87%e0%a4%aa%e0%a4%be%e0%a4%b2-%e0%a4%ad%e0%a4%be/. Accessed on 11 November 2019.

129 Ministry of External Affairs, Government of India. "Transcript of Weekly Media Briefing by Official Spokesperson (November 7, 2019)". https://mea.gov.

in/media-briefings.htm?dtl/32019/Transcript_of_Weekly_Media_Briefing_by_Official_Spokesperson_November_7_2019. Accessed on 12 November 2019.

130 Ministry of Foreign Affairs, Government of Nepal. "Nepal-China Relations". https://mofa.gov.np/nepal-china-relations/. Accessed on 12 October 2019.

131 Ibid.

132 Ministry of Foreign Affairs, Government of Nepal. "Nepal-China Relations". https://mofa.gov.np/nepal-china-relations/. Accessed on 18 November 2019.

133 Stobdan, P. (2017, 14 July). "India's Real Problem Lies in Its Bhutan Policy, Not the Border". *The Wire*. https://thewire.in/diplomacy/india-china-doklam-real-problem-bhutan. Accessed on 2 April 2018.

134 Phuntsho, Karma. (2013). *The History of Bhutan*. Noida: Random House, p. 89.

135 Robinson, Francis. (1989). *The Cambridge Encyclopedia of India, Pakistan, Bangladesh, Sri Lanka, Nepal, Bhutan and The Maldives*. Cambridge: Cambridge University Press.

136 Kohli, Manorama. (1993). *From Dependency to Interdependence: A Study of Indo-Bhutan Relations*. New Delhi: Vikas Publishing House Pvt Ltd.

137 Robinson, Francis. (1989). *The Cambridge Encyclopedia of India, Pakistan, Bangladesh, Sri Lanka, Nepal, Bhutan and The Maldives*. Cambridge: Cambridge University Press, p. 164.

138 "India-Bhutan Relations". Ministry of External Affairs, Government of India. http://mea.gov.in/Portal/ForeignRelation/Bhutan-February-2012. Accessed on 27 March 2018.

139 Ghosh, Shubam. (2014, 18 June). "Understanding India -Bhutan Relations". *One India*. www.oneindia.com/feature/understanding-india-bhutan-relations-1467521.html. Accessed on 27 March 2018.

140 Ibid.

141 Ibid.

142 Ministry of External Affairs, Government of India. "India-Bhutan Relations". https://mea.gov.in/Portal/ForeignRelation/Bhutan-February-2012.pdf. Accessed on 17 April 2018.

143 Ministry of External Affairs, Government of India. "Treaty of Perpetual Peace and Friendship". https://mea.gov.in/bilateral-documents.htm?dtl/5242/treaty+or+perpetual+p. Accessed on 18 January 2019.

144 Ibid.

145 Ministry of External Affairs, Government of India. "India-Bhutan Friendship Treaty". https://mea.gov.in/Images/pdf/india-bhutan-treaty-07.pdf. Accessed on 27 March 2018.

146 Ibid.

147 Ibid.

148 "Agreement on Trade, Commerce and Transit Between the Royal Government of Bhutan and the Government of the Republic of India". Ministry of Economic Affairs. www.moea.gov.bt/wp-content/uploads/2017/07/2016-En-Agreement-on-Trade-and-Commerce.pdf. Accessed on 15 July 2019.

149 Mitra, Devirupa. (2017, 27 April). "With Bhutan Out, Modi's Plan for South Asian Motor Vehicle Movement Is Down to Three Countries". *The Wire*. https://thewire.in/diplomacy/bbin-agreement-without-bhutan. Accessed on 28 June 2019.

150 From 1945 to 1971 it was Republic of China which was member of the United Nations.

151 Mathou, Thierry. "Bhutan China Relations: Towards a New Step in Himalayan Politics". *Bhutan Studies*. www.bhutanstudies.org.bt/publicationFiles/ConferenceProceedings/SpiderAndPiglet/19- Spdr&Pglt.pdf. Accessed on 17 September 2018.

152 Ibid.
153 Ibid.
154 Ibid.
155 Ibid.
156 Ministry of Foreign Affairs, Royal Government of Bhutan. "Press Release". www.mfa.gov.bt/?p=3522. Accessed on 17 September 2018.
157 Ministry of Foreign Affairs, Royal Government of Bhutan. (2017, 29 June). "Press Release". www.mfa.gov.bt/?p=4799. Accessed on 3 September 2017.
158 Ministry of Foreign Affairs of the People's Republic of China. "Assistant Foreign Minister Kong Xuanyou Visits Bhutan, 2018/07/24". www.fmprc.gov.cn/mfa_eng/wjbxw/t1580397.shtml. Accessed on 3 September 2018.
159 Ibid.
160 Ibid.
161 Roy, Shubhajit. (2018, July). "China Minister Arrives in Bhutan on First Top-Level Visit After Doklam". *The Indian Express*. https://indianexpress.com/article/india/china-minister-arrives-in-bhutan-on-first-top-level-visit-after-doklam-5272450/. Accessed on 26 July 2018.
162 Ranjan, Amit. (2019, December). "China's Infrastructure Projects in South Asia: An Appraisal". *Contemporary Chinese Political Economy and Strategic Relations; Kaohsiung*, Vol. 5, No. 3, 1079–1110.

2 River water disputes between India and Pakistan[1]

In 1947 Pakistan was born as a result of the Partition of British India. Since then India and Pakistan have fought three total wars, one limited war, and have been in tensed relationships, with some phases of good relations. Water has never been a major or chief reason to trigger any of those wars or tensions; however, the situation is changing now. The widening of the demand-supply gap of water due to increasing population in both countries and the impact of climate change which affects precipitation are creating competition between them to utilise more waters from the transboundary rivers. They are also competing and contesting over Hydroelectric Power (HEP) projects on their shared rivers. Both water sharing and HEP projects have made the transboundary river waters as a resource of disputes between India and Pakistan. It is forecasted that this dispute is going to escalate further in the future, as demands for water are going to rise in the region where their shared rivers flow.

More than the supply-demand gap, India-Pakistan water disputes, mainly, reflects their bilateral relationships. With the coming of the government under the Hindu nationalist party-Bhartiya Janata Party – led by Narendra Modi in India in 2014, India-Pakistan relations have plummeted further. Modi's government has cleared its intentions over the transboundary rivers with Pakistan in 2016 and again in 2019. This chapter looks at the India-Pakistan water relations since the Partition of British India. It discusses the partition of the irrigation system while partitioning British India and then in 1960 shared rivers were divided. It further examines the water-related politics between the two countries.

Partition of a single irrigation system

The northern part of India and large parts of Pakistan are fed by the Indus Rivers System (IRS), which comprises Indus, Jhelum, Chenab, Ravi Beas, and Sutlej, as well as its extended tributaries, Kabul and Khurram, which rise in Afghanistan. These rivers along with many small tributaries have fed this region through centuries. They have helped in the settlement of human beings and the beginning of agricultural practices on the banks of

IRS, which resulted in the beginning of the Indus Valley Civilisation (3300 to 1300 BC), one of the most developed civilisations in the ancient world. The Indus Valley Civilisation was born, settled, and collapsed on the banks of the River Indus. Even the word "Hindu" (a religious group) has been derived from the River Indus.[2]

In ancient India, to promote agriculture activity small size irrigation structures were built by the rulers, but the foundation of a large irrigation and canal system was, chiefly, set up during the Mughal period (1526–1857). A large number of canals were constructed to facilitate agricultural activities in the areas ruled by the Mughal kings. In Punjab, a small system of canals was brought into existence in the Upper Bari Doab. One of the best known was the "Shahnahr", excavated during the reign of Shahjahan. It took off from the River Ravi at Rajpur (or Shahpur) and carried water up to Lahor (Lahore), which was at a distance of about 37 *kurohs*, or 84 miles.[3] These canal systems helped the Mughals to gain taxes out of agricultural produce and allied activities in the region.

Although the Mughals led the foundation and set up irrigation system in the IRS catchment area, the revolutionary transformation of agriculture and canal system was brought during the British colonial rule in India. After their annexation of Sindh in 1842 and Punjab in 1849, the British colonial rulers constructed a series of canals.[4] Through those canals in Punjab, six million acres of desert were transformed into one of the richest agricultural regions in Asia.[5] Elizabeth Whitcombe finds that introduction of a canal irrigation system in semi-arid plains of north-western British India brought negative ecological consequences in the form of water logging, salinization, and malaria and destroyed traditional wells.[6] She found them "costly experiment".[7] Ian Stone, however, observes that the canal irrigation system had positive impact, as it changed the agriculture in north India.[8] Such transformation in irrigation system also overhauled the agrarian structure and rural set up in Punjab.[9]

After spread of irrigation system in the IRS region, agriculture was commercialised and farmers were encouraged to grow "cash crops" (mainly indigo, cotton, etc.) instead of food grains[10], which was a reason for intermittent famines and starvation deaths in India, including in water rich areas. Besides economic gains, the construction of canals was also related to the political imperatives of state-building in the Indus Basin region. For the British, as much as for the earlier Indus Basin states, the link between canal building, agricultural settlement, and political control was central to the construction of state power.[11] As Sir Charles Aitcheson maintained "It is of the greatest importance to secure for these tracts manly peasantry capable of self-support and of loyal and law-abiding disposition".[12] In these areas canal colonies were situated in tracts designated as crown waste land. Since the owner of the land was the state, it controlled the canal system, the water source and agriculture depended on the will of the ruling authority.[13] The state distributed the land in canal colonies to the loyalist castes and loyal soldiers. In the process a class system was formed where some got land (on lease, not on hereditary basis), while others were made a part of it to do menial works. This structure helped to bring prosperity in the region, though a lopsided one.[14]

In 1947, this physically interdependent and colonial state constructed loyal class-based irrigation system in the IRS catchment region was partitioned between two sovereign countries, who were born to remain in a permanent conflict.[15] As the Partition of British India was "claimed" to be on the basis of religion,[16] the border demarcation was supposed to be on the basis of religious demography; but there were so many interrelated complexities that "other factors" too were taken into account. The terms of reference to the Boundary Commissions (BC) set up to demarcate boundary between India and Pakistan were: "The Boundary Commission is instructed to demarcate the boundaries of the two parts of Bengal on the basis ascertaining the contiguous areas of Muslims and non-Muslims. In doing so, it will also take into account other factors".[17]

In Punjab, which is located in the IRS catchment region, besides, Sir Cyril Radcliffe other members of the BC were: Justice Mehr Chand Mahajan, Justice Teja Singh, Justice Din Mohammad, and Justice Muhammad Munir. In the lawyers dominated Punjab BC, the only technical expert was an Australian geographer, Oskar Spate, who was hired by the Muslim League.[18] Spate was, mainly, recommended by the Ahmadiyya group within the Muslim League to ensure that their religious places become part of Pakistan.[19] However, Spate could not protect Qadian, centre of Ahmadiyya's history, which is in Gurdaspur district of Punjab in India.

There is no clarity on what constituted "other factors". About "other factors", Lucy Chester writes that "at the time of drawing partitioning line between India and Pakistan, in both Punjab and Bengal award, Radcliffe discussed canals, canal head works roads, railways and ports before turning to population factors".[20] She feels that this was aimed to balance the division of religious majorities with the preservation of Punjab's life-giving irrigation system.[21] In some cases, Radcliffe gave more importance to "other factors" than religious demography.[22] Muhammad Munir, one of the judges representing the Muslim League, in Punjab Boundary Commission, independently recalled that "the preservation of the present [1947] irrigation system was an obsession with Sir Cyril".[23]

It is argued by some scholars that some of the decisions, such as of Ferozepur and Gurdaspur, were entirely in consideration of the "other factors".[24] According to Radcliffe's private secretary, Christopher Beaumont "Mountbatten persuaded Radcliffe of the importance of the adverse effects of Bikaner state if Ferozepur went to Pakistan, as well as the possibility of civil war".[25] This was after Maharaja of Bikaner met Mountbatten and argued that award of Ferozepur Tehsil to Pakistan "may gravely prejudice [the] interest of Bikaner State".[26] Finally, Ferozepur and Bikaner became a part of India.

In the Gurdaspur district of Punjab, the Muslim population, according to the 1941 census was: 52.1% in Gurdaspur Tehsil, 55.06% in Batala Tehsil, 51.3% in Shakargarh Tehsil, and 38.8% in Pathankot Tehsil.[27] In the award, except Shakargarh Tehsil all three were allotted to India. It was mainly done to preserve the irrigation system of the Upper Bari Doab Canal which extends from Madhopur in Pathankot tehsil of the district to the western border of the Lahore district. The BC made certain adjustments in the Lahore-Amritsar district boundary to mitigate consequences of the severance.[28] Radcliffe observed:

> I have not found it possible to preserve undivided the irrigation sys-
> tem of the Upper Bari Doab Canal, which extends from Madhopur in
> the Pathankot Tahsil to the western border of the district of Lahore,
> although I have made small adjustments of the Lahore-Amritsar district
> boundary to mitigate some of the consequences of this severance; nor
> can I see any means of preserving under one territorial jurisdiction the
> Mandi Hydro-electric Scheme which supplies power in the districts of
> Kangra, Gurdaspur, Amritsar, Lahore, Jullundar, Ludhiana, Ferozepore,
> Sheikhupura and Lyallpur. I think, it only right to express the hope that,
> where drawing of a boundary line cannot avoid disrupting such unitary
> services as canal irrigation, railways and electric power transmission, a
> solution may be found by agreement between the two States for some
> joint control of what has hitherto been a valuable common service.[29]

Though in Punjab flow of the rivers were used to demarcate boundary between India and Pakistan, Radcliffe was careful to specify that the relevant administrative boundaries, not the course of Ujh, Sutlej, or the Ravi, constitute the new international boundary.[30] On the partition of canals and river system the Chairman of Boundary Commissions report writes:

> The task of delimiting a boundary in Punjab is a difficult one. The
> claims of the respective parties ranged over a wide field of territory,
> but in my judgment the truly debatable ground in the end proved to lie
> in and around the area between the Beas and Sutlej Rivers on the one
> hand, and the River Ravi on the other. The fixing of a Boundary in this
> area was further complicated by the existence of canal systems, so vital
> to the life of Punjab but developed only under the conception of a single
> administration, and of a systems of road and rail communication which
> have been planned in the same way. . . . After weighing to the best of
> my ability such other factors as appeared to me relevant as affecting
> the fundamental basis of contiguous majority areas, I have come to the
> decision set out in Schedule which thus becomes the award of the Com-
> mission. I am conscious that there are legitimate criticisms to be made
> of it; as there are, I think, of any other line that might be chosen.[31]

Further Radcliffe writes:

> I have hesitated long over those not inconsiderable areas east of the Sut-
> lej River and in the angle of the Beas and Sutlej Rivers in which Muslim
> majorities are found. But on the whole I have come to the conclusion
> that it would be in the true interest of neither State to extend the terri-
> tories of the West Punjab to a strip on the far side of the Sutlej and that
> there are factors such as the disruption of railway communications and
> water systems that ought in this instance to displace the primary claims
> of contiguous majorities. But I must call attention to the fact that the
> Dipalpur Canal which serves areas in the West Punjab, takes off from
> the Ferozepur headworks and I find it difficult to envisage a satisfac-
> tory demarcation of boundary at this point that is not accompanied by
> some arrangement for joint control of the intake of the different canals
> dependent on these headworks.[32]

Despite Radcliffe's support for having some joint control or cooperative
mechanisms, the political leadership of India and Pakistan were not ready
for it. To his offer of such possibility, Mohammad Ali Jinnah, leader of Muslim League, replied that "he would rather have Pakistan deserts than fertile
fields watered by the courtesy of the Hindus"[33] while first Prime Minister
of India Pandit Jawaharlal Nehru "curtly informed [Radcliffe] that what
India did with India's rivers was India's affair".[34] The territorial demarcation of Ferozepur and Gurdaspur are still being debated and looked at by
some Pakistani scholars and commentators as a ploy to smother the idea of
Pakistan during the early days.[35]

Post-Partition water disputes

Soon after the Partition of the British India in 1947, the water disputes
between Indian (East) Punjab and Pakistan side of (West) Punjab started.
To address it, a "Standstill Agreement" was signed between them in December 1947 to maintain pre-partition level allocation of water from India to
Pakistan. That agreement was expired on 31 March 1948, the same day
on which the Arbitral Tribunal went out of existence. As a result, on 1
April 1948, East Punjab cut off supplies on the canals to West Punjab.[36] and
discontinued the delivery of water from the Ferozepur headwork to Dipal-
pur canal and to the main branches of the Upper Bari Doab Canal.[37]

India said that the supply has been cut because of non-availability of
waters in the canal system. It is assumed that, technically, had the Arbitral
Tribunal continued the disputes over the canal supplies could have referred
to it.[38] As the Eastern Punjab was not ready to discharge water to Western
Punjab, to find a solution the Inter-Dominion Conference was held in New
Delhi on 3–4 May 1948. Deadlock continued during the conference and it

was broken only after the matter was looked at by Nehru. Eventually, after two days of conference and Nehru's interference, Inter Dominion Agreement was signed between India and Pakistan.[39]

During the conference, both countries adamantly claimed themselves as owner of the shared water resources of Punjab. Pakistan claimed that it had inherent and prescriptive rights over the waters of the canal in dispute, and East Punjab counter claimed that after the partition of India in 1947, it alone had an undisputed right over the water of the canal which lay in its territory.[40] According to May 4 agreement, Pakistan had to cancel its claim in regard to the ownership of the canal; in return East Punjab assured that the water supply would continue for a specific period.[41] Also, Pakistan had agreed to pay seigniorage charges to India for supply of water from East Punjab.[42] On charges for water, the Inter Dominion Agreement said:[43]

1   The West Punjab Government agreed to deposit immediately in the Reserve Bank such an ad hoc sum as may be specified by the Prime Minister of India. Out of the deposited amount, the Government agrees to immediately transfer some money to East Punjab Government over which there is no dispute.
2   After an examination by each party of the legal issues, of the method of estimating the cost of water to be supplied by the East Punjab Government, and of the technical survey of water resources and the means of using them for supply to these canals, the two Governments agree that further meetings between their representatives should take place.

A year after, in a note dated 16 June 1949, Pakistan expressed its displeasure with the May 4 agreement. It talked about "equitable apportionment of all common waters," and suggested for using the services of International Court of Justice.[44] On this, India suggested rather that a commission of judges from each side looking at the water issue and try to resolve their differences before turning the problem over to a third party. Such stalemate remained through 1950.[45] Again, in 1951 the Pakistani leaders repudiated the 4 May 1948 as unfair.[46]

Moreover, soon after agreed May 4 agreement faced another hurdle: it could not be recognised and registered as an international agreement between two sovereign countries. The reason for it was, as India and Pakistan were members of the British Commonwealth, they were not considered as a "foreign" territory to each other.[47] Finally, that issue was sorted out by the United Nations (UN), which recognised both of them as separate countries, so "foreign" to each other. And in May 1950 the UN duly registered not only the May 4 agreement on canal water but also several India-Pakistan agreements dating from between 1948 and 1950 on monetary arrangements, banking, and foreign exchange transactions.[48]

More than Punjab, water issues between India and Pakistan are complicated in J&K through which the IRS is considered as a lifeline of Pakistan and passes through. Besides religion, one of the reasons why Pakistan desires for the Kashmir valley is its waters. This had been substantiated by a statement in 1957, by then Pakistani Prime Minister, Hussain Suharwardy "There are as you know six rivers (in the Indus Basin). Most of them rise in Kashmir. One of the reasons, why, therefore, that Kashmir is so important for us is this water, which irrigate our lands".[49] Even the first military ruler of Pakistan, General Ayub Khan, in his autobiography *Friends Not Masters*, has cited similar reason to explain why Kashmir valley is important for Pakistan. Citing the British Foreign Office source of 1949, Daniel Haines writes "Pakistan might not insist on a plebiscite if an alternative settlement guaranteed the accession of Kashmir up to and beyond the far bank of the Chenab. Anything less would seem to be a surrender to Indian interests".[50] This so-called Chenab formula would re-emerge in the future also.

After 1947, Pakistan had constructed a couple of link canals which ensured for her 50% of the supplies she received from the eastern rivers.[51] Pakistan also insisted on the continuance of the historic supplies as ever before. At that time, India agreed to pay a certain sum of money if it were a part of a final settlement of the dispute. However, no agreement was reached at that time because for Pakistan replacement work could not be separated from development work.[52] Pakistan insisted on third party mediation and raised the issue at various international forums including in the United Nations Security Council on 16 December 1952.[53] On the other hand, India was against any third party involvement in dispute settlement and urged that Inter-Dominion Agreement be made permanent.[54]

In 1951, on the invitation of Nehru, David E. Lilienthal, former chairman of Tennessee Valley Authority and the United States Atomic Energy Commission, came to India as prime minister's guest and paid a visit to the Indus basin. Later, he wrote an article published in the popular American magazine *Colliers* where he made the following proposals:[55] (a) The Indus basin water resources are sufficient to continue all existing uses and to meet the further needs of both countries for water from that source. (b) The water resources of the Indus basin should be cooperatively developed and used in such a manner as to promote most effectively the economic development of the Indus basin viewed as a unit. (c) The problem of development and use of the Indus basin water resources should be solved on a functional and not on a political plane without relation to the past negotiations and past claims independently of political issues. On the issue of politicisation of the water disputes between India and Pakistan, Lilienthal wrote:[56]

> Planning of the water resources (and the accompanying potential electric power) of the Indus Basin is nothing new; it was for generations largely a function of British-trained Indian engineers of the state of Punjab.

They saw the river basin as a unit, as it is in nature. Then partition, a politico-religious instrument, fell like an axe, and colleagues who had worked together all their lives, elbow to elbow, separated because they were Hindu and Sikh or Moslem. Partition did not repeal engineering or professional principles among these engineers; it merely made them secondary for a time to politics and emotion.

This article influenced Eugene Black, the then president of the World Bank, to offer the help of the World Bank to resolve water disputes over the IRS, which was reluctantly accepted by India. The World Bank made it clear that it would not adjudicate the dispute but work as a conduit to agreement. Second, it made a difference between the "functional" and "political" aspects of the Indus dispute, and maintained that it would only look into the "functional" aspects and not "political" aspects of the disputes.[57] Afterwards, in August 1951, Black invited prime ministers of India and Pakistan for a meeting in Washington. Both accepted the invitation and agreed on outline of essential principles.[58] In May 1952, the first meeting of the working party constituting engineers from India and Pakistan was held in Washington.[59]

It is being argued by some former Indian diplomats and scholars that the USA played an important role in the signing of the IWT because of its security concerns in South Asia, especially to protect Pakistan's interests.[60] During the final days of negotiation, the then president of the USA, Dwight D. Eisenhower, wrote a letter to the Nehru and Pakistan's President Iskandar Mirza and discussed about the IRS waters and treaty. It is believed that India's economic situation at that time made Nehru to sign the IWT.[61]

During negotiations Pakistan desired to confine the Working Party to a consideration of what it called the three "common" rivers, the Ravi, the Beas, and the Sutlej (and to keep the Chenab, the Jhelum and the main Indus outside the scope of discussions) for the purpose of determining the "surplus" waters, i.e., waters over and above what Pakistan regarded as "existing" uses and then to proceed to distribution of only such "surplus" between the two countries. India, on the other hand, felt that the Working Party had to determine the *total* waters between the two countries, taking into account the *total* requirement of each.[62] The Working Party accepted India's point and included all rivers, which constitute the IRS for talk.

After eight years of negotiations, mediated by the World Bank, India and Pakistan signed the IWT on 19 September 1960 at Karachi. Nehru visited Pakistan to sign the IWT. Before going to Karachi, when the framework of the IWT was mutually accepted, explaining some of its features, Nehru, answering a question in the Indian Parliament said:

Pakistan should build these replacement works, presumably in ten years' time, and during these ten years' time, and during these ten years we supply water to them, though in a progressively diminishing degree. In building these works, Pakistan is going to be helped by us financially to the extent that we are going to deprive her of water that she has been getting so far. In effect, however, Pakistan is going to build on a much bigger scale with the help of a number of countries and the World Bank. Large sums of money are going to be given to Pakistan by the World Bank and by a number of other countries. But that has nothing to do with our agreement. We are going to make an *ad hoc* contribution spread over ten years.[63]

He added

It has taken a long time to decide how much water we are to give during the transition period of ten years and in what form the payment should be made. The ten year period began in April 1, 1960, the date on which the treaty came into effect, and it can be extended by a further period of three years at Pakistan's request. The extension is subject to a reduction in our contribution by 5% in the first year, 10% over two years and by 16% over the three years. The ten year period is to be roughly divided into two phases, 1960–66 and 1966–70. The water to be supplied by India to Pakistan from the eastern rivers during the transition period is to be of a diminishing scale. India will have no responsibility for their canals, etc.[64]

While signing the treaty Nehru hoped that the agreement would bring prosperity to peasants, and peace, friendship, and goodwill between India and Pakistan.[65] Nehru said:

This is indeed a unique occasion and a memorable day, memorable in many ways, memorable certainly in the fact that a very difficult and complicated problem which has troubled India and Pakistan for many years has been satisfactorily solved. It is also memorable because it is an outstanding example of a co-operative endeavor among our two countries as well as other countries and International Bank.[66]

Highlighting the benefits of the IWT, Nehru further stated:

This settlement is memorable because it will bring assurance of relief to large numbers of people – farmers and others – in Pakistan and India. . . . By this arrangement we have tried to utilize to the best advantage the waters of the Indus River system. These waters have flowed down for ages, the greater part going to the sea without being utilized. This is a happy occasion for all of us. The actual material benefits which will arise from this are great. But even greater than these material benefits are the psychological and emotional benefits. This treaty,

Mr. President (addressing Ayub Khan), is a happy symbol of larger co-operation between your country and mine. I should like to express my deep gratitude to the International Bank and to all those who labored within Pakistan, in India and in other friendly countries, and to all who have come to our assistance in this matter and generously made contributions towards solving this problem . . . hope that this will bring prosperity to a large number of people on both sides and will increase the goodwill and friendship between India and Pakistan.[67]

The Vice President of the World Bank, W.A. B. Illitif, said "*Insha Allah* (God Wiling) this treaty would end for all time to come the Indus water dispute between two countries".[68] R.H.M. Thomson, British parliamentary undersecretary, described the treaty as "significant not only to India, Pakistan and Commonwealth but to the whole world".[69] William M. Rountree, the US ambassador to Pakistan represented the six donor countries.[70] With the coming into effect of the IWT, the Inter Dominion Agreement of 1948 became ineffective from 1 April 1960.[71]

The IWT allocates the three western rivers to Pakistan – Indus, Jhelum, and Chenab – barring some limited uses for India in Jammu & Kashmir. India got the entire waters from other three rivers (Ravi, Beas, and Sutlej), less some minor irrigation uses for Pakistan from four nullahs that joins the River Ravi without making any claims.[72] India is allowed 3.60 Million Acre Feet (MAF) of storage (0.40 MAF on Indus, 1.50 MAF on the Jhelum, and 1.70 MAF on the Chenab). Sector wise allocation is: 2.85 MAF for conservation storage (divided into 1.25 MAF for "general storage" and 1.60 for "power storage") and an additional 0.75 MAF for "flood storage".[73] In a nutshell, of the total average annual flow of about 168 Million Acre Feet (MAF) in the Indus Rivers System (IRS), India can utilise 33 MAF of waters from the eastern rivers and 3.60 MAF from western rivers. On eastern rivers, out of the 33 MAF, India uses about 94 to 95% of its share; the remaining waters flow into Pakistan.

Compromises and adjustments are needed to make any treaty accepted by the parties in dispute, and the IWT was not an aberration. Indian media, the political class, and scholars felt that India had made compromises while signing the IWT. *The Hindu* mentioned some of those adjustments and compromises. *First,* India's right to the use of the western rivers as they flow through the Indian side of J&K to Pakistan has been effectively cut excepting for minor use in J&K. *Second,* it has made India pay for the replacement work. *Third,* India's payment has been separated from the other financial disputes pending between the two countries. *Fourth,* it has been conceded that replacement is as important as development. Whatever additional irrigation potential is developed in Pakistan will be shared between replacement and development. The transition period of ten years has been worked out on this basis. *Fifth,* while India got an additional 3 MAF of water, Pakistan

Figure 2.1 River Chenab in Kashtiwar, Kashmir valley
Source: Photo courtesy of Shabbir Raina

got altogether 35 MAF.[74] After this experience of mediation over the IWT, India shied away from accepting it as a means to resolve or address its other transboundary river water disputes.

In the final reckoning, Pakistan got 80% of the IRS and India 20%. Due to it and many other reasons former diplomat Rajiv Dogra calls the IWT as "the most generous treaty drawn in favour of downstream country".[75] India has limited rights to the western rivers and cannot undertake projects on those rivers without providing all the details to Pakistan and dealing with Pakistan's objections. The big question is Why did India put itself in that position?[76] The answer is that if Pakistan got the near exclusive allocation of the three western rivers, India for its part got the eastern rivers. This was important from the point of view of the Indian negotiators, because the water needs of Punjab and Rajasthan weighed heavily on them in seeking an adequate allocation of IRS waters for India.[77]

The demand of Kutch (in Indian Gujarat), which used to fall into a catchment area of River Indus, decades back, was not taken into consideration despite many petitions, arguing about their historical claim on its water, sent by the prominent Kutchi leaders, in 1950s, to the Ministry of Irrigation and Power, Government of India.[78] At the time of negotiations, the government

Figure 2.2 River Jhelum, Zirpora Bijbehara, Kashmir valley
Source: Photo courtesy of Shabbir Raina

of India took up the case of Rajasthan where water used to be supplied through canal system. On this issue, Pandit Nehru said:

> According to present plans, the Rajasthan Canal will be ready to carry some irrigation water up to 1,200 cusecs in 1961, 2, 100 cusecs in 1962 and 3,000 cusecs in 1963. Thereafter it is proposed to enlarge the capacity in such a way that by 1970 the Canal would be developed to 18,500 cusecs.[79]

Years after the IWT was signed, people from the Indian side of Kashmir expressed their ire against the IWT. On 3 April 2002, the Jammu & Kashmir

Figure 2.3 River Sutlej, Rampur, Himachal Pradesh
Source: Photo courtesy of Vishwamitra Negi

Legislative Assembly, cutting across party affiliations, called for a review of the treaty. The State government has been contending that, in spite of untapped hydroelectric potential of 15,000 MW, the state has been suffering from acute power deficiency due to restrictions put on the use of its rivers by the Indus Treaty. They claim that their interests were not taken into consideration and their views were not taken into account while signing the treaty. Baglihar's judgment is considered as blessing in disguise for some people from the Jammu & Kashmir.[80] This is because the run-of-the-river dam can produce 450 MW of hydroelectricity to be utilised by the people from the state. Also, the state is entitled to get royalty from the power producing companies. In 2015 the Minister of State for Power, Mohammad Ashraf Mir, in a written reply to the question of Bashir Ahmed Veeri, informed the members that Indian Rupees 2472.37 crores (US $32.4 billion) were received from National Hydroelectric Power Corporation (NHPC) Limited as water usage charges from year 2010–11 to 2015.[81] Enthusiastic about the prospects of financial gain, a cabinet subcommittee was constituted to look into various issues associated with transfer of Power projects to the State, which recommended seeking return of Salal, Uri-I, and Dulhasti Hydroelectricity Projects to the State apart from those transferred to NHPC vide MoU of July 2007.[82]

Contrary to such positive perceptions, J&K reel under power cut during harsh winters. Kashmiris feel that their resources are not under their control. There are 21 state-owned projects with an estimated capacity to generate 1,211.96 megawatts of power besides seven **projects under central government which can produce** 2,009 megawatts of power. Despite so much production, local demands are not being fulfilled. In 2016 the peak demand increased by 8% and about three lakhs (300,000) home remained non-electrified.[83] This treatment, many in Kashmir valley feel, is discriminatory and this is why the NHPC has earned local sobriquet of the new "East India Company"[84] which is engaged in exploiting local resources for the benefits of "outsiders". Such feeling has been strong due to political alienation of a number of people from valley while living under the Armed Forces Special Powers Act from many years. The revocation of special provisions under article 370 of the Indian constitution in 2019 may alienate more Kashmiris.[85] Even militant groups have played their role in the gradual increase in the number of disgruntled and alienated individuals in the Kashmir valley. Such groups have attracted the youths to join them and engage in militant activities.[86] After scrapping of the special status of J&K, a report prepared by the Indian security agencies says that the number of youths joining militancy has gone down.[87] However, the report has not been scrutinised and vetted by the non-State actors and independent institutes.

Like in India, in Pakistan too grievances remain high against the IWT. Pakistani scholars feel that the IWT's provisions were set up and are highly in favour of India. It gives India control over Pakistan's water resources, they feel.[88] At provincial level, Sindh has accused Punjab of sacrificing the water interests of other provinces during the IWT negotiations. One basis for such arguments is that in 1955 a Sindhi, Saleh Qureshi, was removed from the Pakistani negotiation team when IWT discussions were in final stage.[89]

Despite all grievances, the IWT is still intact and provisions under it, more or less, have been abided by the two signatories. Even three wars (1965 and 1971 and 1999) have not disrupted this treaty. The great example of cooperation on this treaty is: even in the midst of the 1965 war, Indian payments to Pakistan as part of the IWT continued uninterrupted, as did the works of engineers of both countries to control the opening and closing of sluices.[90] During all their wars the two sides have also not directly attacked each other's irrigation facilities, which would cause greater catastrophe to India or Pakistan. The fact that the IWT and the irrigation works survived the crucial test is evidence of their mutual value to each nation.[91] But this does not mean that both sides have accepted the IWT; voices have been raised against the continuation of the IWT.

Growing Water Nationalism

Over the years, as India-Pakistan political tensions have escalated, supply-demand of water gap have further widened, and there has been a rise in demands for electricity, "water nationalism" has been strengthened in the respective countries. An example is in March 2010, when Jamad-ud-Dawa chief Hafiz Saeed in his rallies at Muzaffarabad and Lahore, denounced India's "theft" of waters through "illegal dams" that could trigger nuclear war. Banners in his rallies had slogans like "water or war", "water flows or blood", "Liberate Kashmir to secure water", and "No peace with Indian water aggression".[92]

Unlike in the past, there have been also military attacks on constructions over the shared rivers. For example, in 2018, in what is considered as an attempt to stop the then ongoing construction works on the Kishanganga Hydroelectricity project, 18 shells were fired from across the Line of Control by Pakistan. The aim of the shells was the tunnel which takes the water from the Kishanganga River in Gurez Valley to an underground power station at Bandipora in the Kashmir Valley.[93]

Like Pakistan, in India too "water nationalism" exists. In 2008, after the terrorists attack in Mumbai, voices were raised to abrogate the IWT.[94] Unlike earlier, in 2016, for the first time, a question over the IWT's continuity was raised by the Indian prime minister. After the militant attack on the Indian Army's camp at Uri in J&K, in which about 20 Indian soldiers were killed, Narendra Modi, stated that "Blood and Water cannot flow simultaneously".[95] Thereafter, India did not attend the annual meetings of the Permanent Indus Commission. However, the gap did not extend for a long time period. India attended the Indus commission meeting at Islamabad in 2017. Subsequently, in 2018 the Indus Commissions held meetings in New Delhi. There the two sides held technical decisions on various hydroelectric projects, including Pakal Dul (1000 MW) and Lower Kalnal (48 MW). Pakistan had raised objections over the two projects. The two countries also agreed to undertake the IWT's mandated tours of Indus Commissioners in Indus basin on both sides.[96] However, India postponed the inspection of Pakal Dul and Lower Kalnal by Pakistani experts which was seen as an impact of the cancellation of agreed talks between the Foreign Ministers of the two countries at the sidelines of the 73rd session of the United Nations General Assembly in September 2018.[97]

In February 2019, after a suicide bomber in Pulwama in the Kashmir valley rammed an explosive-loaded car into a bus carrying the Central Reserve Police Force soldiers in which about 40 soldiers had lost their lives, voices were raised to revoke the IWT. On 21 February 2019, then Union Minister of Water Resources, Nitin Gadkari, said that the Government of India has decided to stop its share of water from the eastern rivers flowing into Pakistan. Besides, he stated, India could even block Pakistan's share of waters.[98] Later he clarified that the decision to curtail Pakistan's share of water would lead to the abrogation of the IWT, hence it needs consideration from the country's top leadership. His ministry, as Gadkari said, is exploring such options.[99]

Earlier, to utilize the waters of the Eastern flowing rivers, India has constructed the Bhakra Dam on Satluj, the Pong and Pandoh Dam on Beas, and the Thein (Ranjitsagar) on Ravi.[100] These storage works, along with Beas-Sutlej Link, Madhopur-Beas Link, Indira Gandhi Nahar Project etc has helped India utilize nearly entire share 95 % of waters of Eastern rivers.[101] However, about 2 MAF of water annually remain unutilized and flow into Pakistan below Madhopur. [102] To stop such flows India has resumed the construction of Shahpurkandi project, constructing the Ujh multipurpose project and has planned for the second Ravi Beas link below Ujh.[103]

India also claimed that 0.53 MAF waters of the eastern rivers had been stopped. To it, Pakistan's former federal secretary for water resources Khawaja Shamail said, "The diversion and optimum use of water of eastern rivers by India is not problematic for us as we are using our own share of water from the western rivers under the Indus Waters Treaty".[104] He added:

> If we see unused water flow of eastern rivers into Pakistan, keeping in view the flood, normal and dry weather, the average amount of water will be around 1 MAF. During floods, normal and dry seasons, it comes to two to three MAF, 1 MAF and 0.50 MAF. The average remains at 1MAF unused water of eastern rivers that enters Pakistan. So, if they have stopped 0.53 MAF unused water of their share from flowing into Pakistan, we have no problem.[105]

However, in Pakistan the eastern flowing rivers have their significant contribution, though in parts of the country. River Sutlej enters into Pakistan at Kasur in Punjab while Ravi enters the country near Kartarpur Sahib and flows on the northern side of Lahore. They support irrigation, industrial, and domestic needs of a section of population.

In recent years, as India-Pakistan tensions have been escalated, support for revocation of the IWT has been on rise. Many of the Hindi, English, and regional news channels, mainly from India, have repeatedly debated the possibility of revocation of the IWT. However, the IWT is difficult to be scrapped; it does not have unilateral exit clause, though it can be modified from time to time. Article XII (3) of the IWT says that "The provisions of this Treaty may from time to time be modified by a duly ratified treaty concluded for that purpose between the two Governments".[106] Then 12 (4) says "The provisions of this Treaty, or, the provisions of this Treaty as modified under the provisions of Paragraph (3), shall continue in force until terminated by a duly ratified treaty concluded for that purpose between the two Governments".[107]

Technically, under the Vienna Convention on the Law of Treaties, which was done in 1969 and entered into force on 27 January 1980, there are provisions to sever and withdraw from the treaty in Section 3 (Articles 54 to 64). One of the arguments to scrap the IWT can be made using Article 60 of this treaty which calls for termination or suspension of the treaty because of its breach by the other party. However, it's difficult to prove that the lower riparian breaches the provisions of the IWT.[108] Another, Article 62 of the Vienna Convention mentions the change of circumstances from the time the treaty was concluded to at present as a reason to scrap the treaty. Here India may argue about change in circumstances due to Pakistan-sponsored militancy in its territory. But this would not support India's case on the IWT. Only change in circumstances in the case of the IWT is drying up of the IRS basin. Even the severance of the diplomatic and consular relationships between India and Pakistan, which many commentators argue for, cannot terminate the IWT. Article 63 of the Vienna Convention on the Law of Treaty states:

> The severance of diplomatic or consular relations between parties to a treaty does not affect the legal relations established between them by the treaty except in so far as the existence of diplomatic or consular relations is indispensable for the application of the treaty.[109]

Then there are also internationally accepted water sharing rules. One such rule is Berlin rules on Water Resources which was adopted by the International Law association in August 2004. This law calls for the basin states to manage the shared waters in an equitable, reasonable, and sustainable manner. It also ask them to refrain from harming the other basin's water interests by any of their acts. Article 51 of the Berlin law refrain warring states to attack on each other's water resources. It says (written in verbatim):[110]

1   Combatants shall not, for military purposes or as reprisals, destroy or divert waters, or destroy water installations, if such actions would cause disproportionate suffering to civilians.
2   In no event shall combatants attack, destroy, remove, or render useless waters and water installations indispensable for the health and survival of the civilian population if such actions may be expected to leave the civilian population with such inadequate water as to cause its death from lack of water or force its movement.
3   In recognition of the vital requirements of any party to a conflict in the defense of its national territory against invasion, a party to the conflict may derogate from the prohibitions contained in paragraphs 1 and 2 within such territories under its own control where required by imperative military necessity.
4   In any event, waters and water installations shall enjoy the protection accorded by the principles and rules of international law applicable in war or armed conflict and shall not be used in violation of those principles and rules.

In international politics, rules and laws are set up and violated by the powerful countries. It becomes easy for the powerful countries when the international system is weak while difficult in the case when there is a proper balance of power. In the present international system, India, though a rising power, is not powerful enough to challenge and violate the international laws and treaties it has entered into. As IWT was mediated by the World Bank and counter signed, on behalf of the bank, by W.A.B. ILIFF "for the purposes specified in Articles V and X and Annexures F, G and H", any step towards the direction of IWT's revocation by any parties could be understood by the World Bank as a challenge to its office.

Amidst all such water and HEP projects related disputes, India and Pakistan have also shown cooperation. In 2010, India invited Pakistani experts for HEP project site visit on the IRS Rivers. Again, in 2019, though the bilateral relationship has not transformed, a three-member delegation of Pakistani experts, led by Pakistan's Indus Waters Commissioner Syed Mehr Ali Shah, came to India to complete its general tour of inspection to HEP projects – Pakal Dul, Lower Kalnai, Ratlay (850 MW), and Baglihar Dam on the Chenab basin.[111] During that visit, the Pakistan side invited India to visit Kotri Barrage. Few days before the Pulwama attack of February 2019, India also shared the design data of its three planned run-of-the-river HEP schemes with Pakistan under the provisions of the IWT. These projects include Balti Kalan, Kalaroos, and Tamasha hydropower which are planned to be constructed at Balti Kalan Nullah and Kalaroos Nullah at the Jhelum basin and Tamasha, a sub-tributary of the Indus River, respectively.[112]

Conclusion

This chapter has discussed the history of the canal system in present India and Pakistan and its partition in 1947. It has also discussed how India and Pakistan accepted mediation, negotiated, and signed the IWT in 1960. Despite the existence of IWT, the problems and disputes over the water sharing exists, largely, due to political relationships between India and Pakistan. The IWT does not have unilateral exit clause, but it does permit the possibility of re-negotiation and amendments. However, looking at the present status of India-Pakistan relationships any re-negotiation on the IWT is almost non-thinkable.

As examined in this chapter, the political relationship between India and Pakistan influences their water relations. Partition related memories and its construction are potent reasons for the continuing animosities but also related to them is the religious differences between people living across the border. Tensions have increased further after the rise of militancy in Kashmir valley since 1989, for which India accuses Pakistan. On contrary, Pakistan accuses India of militancy and violence in Balochistan and in other parts of the country. These issues do influence decisions over the water sharing between India and Pakistan. After the abrogation of special status of J&K under Article 370 of the Indian constitution by the Government of India, India and Pakistan have almost stopped any forms of bilateral engagements.

Although the disputes over water shape India-Pakistan water narratives, there have been instances of cooperation between them, as mentioned in this chapter. In the coming years, as demand for water is going to increase in India and Pakistan, water-related disputes may further escalate. There are arguments made by few scholars from both sides in support for cooperation to deal with the imminent crisis due to increasing population and climate change. However, looking at the 71 years of the India-Pakistan relationships one can conclude anything except that the two countries can cooperate in the coming years.

Notes

1 A part of this chapter was originally published as "Disputed Waters: India, Pakistan and the Transboundary Rivers". *Studies in Indian Politics*, Vol. 4, No. 2, pp. 191–205 © 2016. All rights reserved. Reproduced with the permission of the copyright holders and the publishers, Sage Publications India Pvt. Ltd, New Delhi and Centre for the Study of Developing Societies, New Delhi.

2 Thapar, Romila. (2014). *The Past as Present: Forging Contemporary Identities Through History*. New Delhi: Aleph Publication, p. vii.

3 Habib, Irfan. (1963). *The Agrarian System of Mughal India*. Bombay: Oxford University Press, p. 37.

4 Whitcombe, Elizabeth. (1983). "Irrigation and Railways". In Tapan Ray Chaudhuri Dharma Kumar, Irfan Habib & Meghnad Desai (ed.), *The Cambridge Economic History of India*, Vol. 2: c. 1757–c. 1970. Cambridge: Cambridge University Press, pp. 677–761.

5 Talbot, Ian. (2007). "Punjab Under Colonialism: 'Order and Transformation in British India' ". *Journal of Punjab Studies*, Vol. 14, No. 1, pp. 1–10.

6 Whitcombe, Elizabeth. (1983). "Irrigation and Railways". In Tapan Ray Chaudhuri Dharma Kumar, Irfan Habib & Meghnad Desai (ed.), *The Cambridge Economic History of India*, Vol. 2: c. 1757–c. 1970. Cambridge: Cambridge University Press, pp. 677–761.

7 Whitcombe, Elizabeth (1996 edition). "The Environmental Cost of Irrigation in British India: Waterlogging, Salinity and Malaria" in David Arnold and Ramchandra Guha (ed) *Nature, Culture, Imperialism: Essays on the Environmental History of South Asia*. New Delhi: Oxford University Press, 237–259.

8 Stone, Ian (1985) *Canal Irrigation in British India: Perspectives on Technological Change in a Peasant Economy*. Cambridge: Cambridge University Press.

9 Bhattacharya, Neeladri. (2018). *The Great Agrarian Conquest*. Ranikhet: Permanent Black.

10 Jodhka, Surinder. (2004). "Agrarian Structures and Their Transformations". In Das, Veena (ed) *Handbook of Indian Sociology*. New Delhi: Oxford University Press, pp. 365–387.

11 Gilmartin, David. (1994, November). "Scientific Empire and Imperial Science: Colonialism and Irrigation Technology in the Indus Basin". *The Journal of Asian Studies*, Vol. 53, No. 4, pp. 1127–1149.

12 Talbot, Ian. (2007). "Punjab Under Colonialism: 'Order and Transformation in British India' ". *Journal of Punjab Studies*, Vol. 14, No. 1, pp. 1–10.

13 Ali, Imran. (1989). *The Punjab Under Imperialism:1885–1947*. New Delhi: Oxford University Press.
14 Ibid.
15 Wolpert, Stanley. (2011). *India and Pakistan: Continued Conflict or Cooperation*. CA: University of California Press, p. 7.
16 Although it is maintained that the partition was on the basis of religion, many Hindus remained in Pakistan and Muslims in India. Many of those who wanted to migrate but could not because they had no economic means and had no relatives on the other side of the border; some remained because they did not want to leave "their" land. Also, as Schedule Caste Federation (SCF) supported Muslim League there were no attacks on lower castes. Another reason for not attacking them was they were considered to be untouchable so not to be touched even by Muslims, who were converted from the upper caste Hindu. See, Butalia, U. (2007). *The Other Side of Silences: Voices From the Partition of India*. New Delhi: Penguin. In India, faith in the non-communal leadership of Pandit Jawaharlal Nehru made many Muslims not leave "their" land, while in Pakistan the situation became worse soon after an early demise of Mohammad Ali Jinnah in 1948. Leader of SCF and the first law minister of Pakistan, Jogendra Nath Mandal resigned from his position and returned to India. Singh, Nandita. (2019). "When a Pakistan Minister & Jinnah Follower Resigned Over Atrocities Against Hindus". *The Print*. https://theprint.in/politics/when-a-pakistan-minister-jinnah-follower-resigned-over-atrocities-against-hindus/211252/. Accessed on 18 April 2020.
17 Ministry of External Affairs, Government of India. "The Gazette of India Extraordinary Part-I Section 1". www.pib.nic.in/archive/docs/DVD_13/. . . BR/EXT-1950-05-02_1259.p. Accessed on 14 February 2019.
18 Chester, Lucy P. (2009). *Borders and Conflicts in South Asia: The Radcliffe Boundary Committee and Partition of Punjab*. Manchester and New York: Manchester University Press.
19 Spate, Oskar. (1991). *On the Margins of History: From the Punjab to Fiji*. Canberra: National Centre for Development Studies, Research School of Pacific Studies, The Australian National University.
20 Chester, Lucy P. (2009). *Borders and Conflicts in South Asia: The Radcliffe Boundary Committee and Partition of Punjab*. Manchester and New York: Manchester University Press, p. 80.
21 Ibid., p. 74.
22 Ibid.
23 Ibid., p. 80.
24 Michel, Alloys Arthur. (1967). *The Indus Rivers: A Study of the Effects of Partition*. New Haven and London: Yale University Press, p. 178.
25 Chester, Lucy P. (2009). *Borders and Conflicts in South Asia: The Radcliffe Boundary Committee and Partition of Punjab*. Manchester and New York: Manchester University Press, p. 120.
26 Ibid., p. 92.
27 Partition Proceedings Volume VI Commission Papers, Reports of the Members and Awards of the Chairman of the Boundary Commissions. (1950). Superintendent, Government Printing West Bengal Government Press, Alipore, p. 243.
28 Ibid., p. 305.
29 Ibid.
30 Chester, Lucy P. (2009). *Borders and Conflicts in South Asia: The Radcliffe Boundary Committee and Partition of Punjab*. Manchester and New York: Manchester University Press.
31 Partition Proceedings Volume VI Commission papers, Reports of the Members and Awards of the Chairman of the Boundary Commissions. (1950). Superintendent, Government Printing West Bengal Government Press, Alipore, p. 304.

32 Ibid.
33 Cited in Talbot, Ian and Gurharpal Singh. (2009). *The Partition of India*. New Delhi: Cambridge University Press, p. 45.
34 Ibid.
35 Hussain, Ijaz. (2017). *Indus Waters Treaty: Political and Legal Dimensions*. Karachi: Oxford University Press.
36 Michel, Alloys Arthur. (1967). *The Indus Rivers: A Study of the Effects of Partition*. New Haven and London: Yale University Press, p. 7.
37 Salman, M.A. and K. Uprety. (2002). *Conflicts and Cooperation on South Asia's International Rivers: A Legal Perspective*. Washington, DC: The World Bank.
38 Michel, Alloys Arthur. (1967). *The Indus Rivers: A Study of the Effects of Partition*. New Haven and London: Yale University Press.
39 "Sharing of Punjab Water: Indo-Pakistan Agreement". *The Hindu*, 1948, 6 May.
40 Ibid.
41 Ibid.
42 Haines, Daniel. (2014). "Disputed Rivers: Sovereignty, Territory and State-Making in South Asia, 1948–1951". *Geopolitics*, Vol. 19, No. 3, pp. 632–655.
43 Ministry of External Affairs, Government of India "Indus Waters Treaty 1960 "https://mea.gov.in/bilateral-documents.htm?dtl/6439/Indus. Accessed on 15 May 2019.
44 Wolf, Aaron T. and Joshua T. Newton. "Case of Transboundary Dispute Resolution: The Indus Water Treaty". https://transboundarywaters.science.oregonstate.edu/sites/transboundarywaters.science.oregonstate.edu/files/Database/Research-Projects/casestudies/indus.pdf. Accessed on 24 December 2019.
45 Ibid.
46 Haines, Daniel. (2014). "Disputed Rivers: Sovereignty, Territory and State-Making in South Asia, 1948–1951". *Geopolitics*, Vol. 19, No. 3, pp. 632–655.
47 Ibid.
48 Ibid.
49 Cited in Alam, Undala Z. (2002, December). "Questioning the Water War Rationale". *The Geographical Journal*, Vol. 168, No. 6, pp. 341–353.
50 Haines, Daniel. (2017). *Indus Divided: India, Pakistan and The River Basin Dispute*. New Delhi: Penguin Publications, p. 67.
51 "Conclusion of Treaty: Reasons for Long Delays". *The Hindu*, 1960, 21 September.
52 Ibid.
53 Thapliyal, Sangeeta. (1996). "Water and Conflict: The South Asian Scenario". *Strategic Analysis*, New Delhi, October 2006, pp. 1033–1051.
54 Salman, M.A. and K. Uprety. (2002). *Conflicts and Cooperation on South Asia's International Rivers: A Legal Perspective*. Washington, DC: The World Bank.
55 Government of India. (1953). *The Indus Basin Irrigation Water Dispute*. New Delhi: Government of India Press, p. 41.
56 Ibid., p. 37.
57 Salman, M.A. and K. Uprety. (2002). *Conflicts and Cooperation on South Asia's International Rivers: A Legal Perspective*. Washington, DC: The World Bank.
58 Wolf, Aaron T. and Joshua T. Newton. "Case of Transboundary Dispute Resolution: The Indus Water Treaty". https://transboundarywaters.science.oregonstate.edu/sites/transboundarywaters.science.oregonstate.edu/files/Database/Research Projects/casestudies/indus.pdf. Accessed on 24 December 2019.
59 Ibid.
60 The author was informed and told by some former diplomats that this was the reason for the signing of the IWT.
61 Stone, David R. (2010). "The United States and the Negotiation of the Indus Water Treaty". USI. http://usiofindia.org/Article/Print/?pub=Journal&pubno=579&ano=709. Accessed on 16 April 2016.

62 Gulhati, N.D. (1973). *Indus Water Treaty*. New Delhi: Allied Publishers, p. 107.
63 Nehru, Jawaharlal. (1964). *Jawaharlal Nehru's Speeches*, Vol. 4. New Delhi: Publications Division, Ministry of Information and Broadcasting, Government of India, p. 291, September 1957–April 1963.
64 Ibid.
65 "Signing of Indus Water Treaty: End of Dispute for All Time". *The Hindu*, 1960, 21 September from the Archives.
66 Nehru, Jawaharlal. (1964). *Jawaharlal Nehru's Speeches*, Vol. 4. New Delhi: Publications Division, Ministry of Information and Broadcasting, Government of India, p. 292, September 1957–April 1963.
67 Ibid., p. 293.
68 "Signing of Indus Water Treaty: End of Dispute for All Time". *The Hindu*, 1960, 21 September.
69 Ibid.
70 Ibid.
71 Ministry of External Affairs, Government of India "Indus Waters Treaty 1960" https://mea.gov.in/bilateral-documents.htm?dtl/6439/Indus. Accessed on 15 May 2019.
72 Ibid
73 Ibid.
74 "Conclusion of Treaty: Reasons for Long Delays". *The Hindu*, 1960, 21 September.
75 " Boo Discussion on " The longest August: Unflinching Rivalry Between India and Pakistan" *Indian Council of World Affairs*. 5 February 2016. https://www.icwa.in/show_content.php?lang=1&level=2&ls_id=2312&lid=1722.   Accessed on 18 June 2019.
76 Iyer, Ramaswami R. (2005, 15 July). "Indus Treaty: A Different View". *Economic and Political Weekly*, pp. 3140–3144.
77 Ibid.
78 Mehta, Lyla. (2005). *The Politics and Poetics of Water: Naturalizing Scarcity in Western India*. New Delhi: Orient Longman.
79 Nehru, Jawaharlal. (1964). *Jawaharlal Nehru's Speeches*, Vol. 4. New Delhi: Publications Division, Ministry of Information and Broadcasting, Government of India, p. 292, September 1957–April 1963.
80 Warikoo, K. (2005, July–September). "Indus Waters Treaty: View from Kashmir". *Himalayan and Central Asian Studies*, Vol. 9, No. 3, pp. 3–26.
81 *The Economic Times*. (2015, 25 March). "J&K Government Receives Over Rs 2,400 Crore as Water Charges from NHPC". http://articles.economictimes.indiatimes.com/2015-03-25/news/60475007_1_salal-uri-i-water-usage-charges. Accessed on 25 March 2018.
82 Ibid.
83 Naqash, Rayan. (2017, November 15). "Behind the Conspiracy Theories: Why Kashmir Reels Under Power Cut Every Winter". *Scroll.in*. https://scroll.in/article/857433/behind-the-conspiracy-theories-why-kashmir-reels-under-power-cuts-every-winter. Accessed on 28 June 2018.
84 Darbu, Iftikhar A. (2018, 17 August). "Power Starved J&K Is Transfer of Hydro Power Projects From NHPC to State a Solution?" *ORF*. https://www.orfonline.org/expert-speak/43373-power-starved-jk-transfer-hydro-power-projects-nhpc-state-solution/. Accessed on 29 November 2018.
85 Ranjan, Amit. (2019, August 9). "What Does the Article 370 Decision Mean for J&K's Already Troubled Ties with Water?" *The Wire*. https://www.orfonline.org/expert-speak/43373-power-starved-jk-transfer-hydro-power-projects-nhpc-state-solution/. Accessed on 9 August 2019.

86 "191 Kashmiri Youths Joined Militancy in 2018". *The Economic Times*, 2019, 04 February. https://economictimes.indiatimes.com/news/defence/191-kashmiri-youths-joined-militancy-in-2018-official/articleshow/67835737.cms?from=mdr. Accessed on 8 February 2019.

87 "Number of youths joining militancy in Kashmir has gone down since 5 August, says report". *The Print*, 2020, 9 February. https://theprint.in/india/number-of-youths-joining-militancy-in-kashmir-has-gone-down-since-5-august-says-report/362175/. Accessed on 12 March 2020.

88 Hussain, Ijaz. (2017). *Indus Waters Treaty: Political and Legal Dimensions*. Karachi: Oxford University Press.

89 Altaf Memon. (2002). "An Overview of the History and Impacts of the Water Issue in Pakistan". *Pakistan Water Gateway*. http://www.waterinfo.net.pk/sites/default/files/knowledge/An%20Overview%20of%20the%20History%20and%20Impacts%20of%20the%20Water%20Issue%20in%20Pakistan.PDF. Accessed on 18 April 2008.

90 Trembley, Reeta C. and J. Schofield. (2006). "Institutional Causes of the India-Pakistan Rivalry". In Paul, T.V. (ed) *The India-Pakistan Conflict: An Enduring Rivalry*. New Delhi: Cambridge University Press, pp. 225–250.

91 Michel, Alloys Arthur. (1967). *The Indus Rivers: A Study of the Effects of Partition*. New Haven and London: Yale University Press, p. 10.

92 Verghese, B.G. (2010, 12 March). "Do Pakistan's Claim Over the Indus Hold Water? *Indian Express*. https://indianexpress.com/article/opinion/columns/do-pakistans-claims-over-the-indus-hold-water/. Accessed on 12 January 2016.

93 Subraminan, N. (2018, 23 May). [In Kishanganga Dam Security, More Than Pakistan Shelling, Sabotage a Concern] *The Indian Express*. https://indianexpress.com/article/explained/kishanganga-hydel-electricity-dam-security-india-pakistan-army-pakistan-shelling-5187314/. Accessed on 26 May 2018.

94 Menon, M.S. (2009, 30 January). "Withdraw from Indus Treaty". *The Tribune*. www.tribuneindia.com/2009/20090130/edit.htm#7. Accessed on 12 December 2017.

95 "Blood and Water Can't Flow Simultaneously". PM Narendra Modi Gets Tough on Indus Treaty] *The Times of India*, 2016, 27 September. https://timesofindia.indiatimes.com/india/Blood-and-water-cant-flow-together-PM-Narendra-Modi-gets-tough-on-Indus-treaty/articleshow/54534135.cms. Accessed on 20 August 2018.

96 Ministry of External Affairs, Government of India. (2018). [115th Meeting of the India-Pakistan Permanent Indus Commission (PIC)] Accessed on 17 August 2018.

97 Hasnain, K. (2018, 26 September). [India Reneges on Deal to Get Power Projects Inspected] *Dawn*. www.dawn.com/news/1435049/india-reneges-on-deal-to-get-power-projects-inspected. Accessed on 26 September 2018.

98 "India Could Block Even Pakistan's Share of Waters: Nitin Gadkari". *The Times of India*, 2019, 23 February. https://timesofindia.indiatimes.com/india/india-could-block-even-pakistans-share-of-indus-waters-nitin-gadkari/articleshow/68120069.cms. Accessed on 24 February 2019.

99 Ibid.

100 Ministry of Water Resources, River Development & Ganga Rejuvenation, Press Information Bureau (2019, 22 February) "Indus Waters Treaty 1960 : Present Status of Development in India". https://pib.gov.in/PressReleseDetail.aspx?PRID=1565906. Accessed on 24 February 2019.

101 Ibid.

102 Ibid.

103 Ibid.

104 Khalid Hasnain. "Indian Blockade of Eastern Rivers Won't Affect Pakistan". *Dawn*, 2019, 12 March. https://www.dawn.com/news/1469046. Accessed on 13 March 2019.
105 Ibid.
106 Ministry of External Affairs, Government of India. "Indus Waters Treaty, 1960". https://mea.gov.in/bilateral-documents.htm?dtl/6439/Indus. Accessed on 12 July 2019.
107 Ibid.
108 "Vienna Convention on the Law of Treaties 1969". United Nations. http://legal.un.org/ilc/texts/instruments/english/conventions/1_1_1969.pdf. Accessed on 22 July 2018.
109 Ibid., p. 22.
110 International Law Association, Berlin Conference. (2004). "Water Resources Law". http://www.cawater-info.net/library/eng/l/berlin_rules.pdf. Accessed on 18 February 2017.
111 Hasnain, Khalid. (2019, 12 March). "Indian Blockade of Eastern Rivers Won't Affect Pakistan". *Dawn*. www.dawn.com/news/1469046. Accessed on 18 August 2019.
112 Ibid.

3   River water disputes between India and Bangladesh

Contrary to India's relationship with Pakistan, Bangladesh is regarded as a close friend, even though several problems, as mentioned in Chapter 1, remain between the two countries. During its East Pakistan (later Bangladesh) days their political relationships had affected the water-related issues but after Bangladesh was liberated in 1971 several such issues have been resolved between them. India and Bangladesh share 54 rivers of which the Ganges, Brahmaputra, and Meghna (GBM) form a basin whose total catchment area is 1.75 million square kilometres of which Bangladesh accounts for 7%, Bhutan 3%, India 63%, Nepal 9%, and Tibet (China) 19%.[1] Some of these are border rivers which often create confusions over how to demarcate the boundary between India and Bangladesh. Out of a total border length of about 4096 kilometres, about 1116 kilometres are riverine, and 2,980 are land.[2]

To settle their water disputes over the River Ganges, in 1996 India and Bangladesh signed a treaty to share its waters. There are certain issues between the two countries over the rivers other than the GBM, but the notable one is sharing of waters from River Teesta. In 2011 the two governments accepted new formula to share Teesta waters, but the Indian state of West Bengal is not yet ready to release the agreed amount of water to Bangladesh. This chapter discusses the river water disputes between India and Bangladesh whose roots like in the case of India-Pakistan water disputes can be located to the Partition of British India in 1947.

Partition and India-Bangladesh water disputes

The Partition of Bengal like that of Punjab disturbed the river systems of the region. As the terms of reference of the Boundary Commission (BC) allowed it to consider "other factors" while demarcating boundaries between the two Bengals, Cyril Radcliffe used it to divide the territories. The Bengal Boundary Commission, under Radcliffe, was constituted on 30 June 1947 and had four members: Justice Bijan Kumar Mukherjea, Justice C.G. Biswas, Justice

Abu Saleh Mohamed Akram, and Justice S.A. Rahman. Largely, the BC had seven questions before it while dividing Bengal:[3]

(i)   To which State was the City of Calcutta to be assigned, or was it possible to adopt any method of dividing the City between the two States?

(ii)   If the City of Calcutta (now Kolkata) must be assigned as a whole to one or the other States, what were its indispensable claims to the control of territory, such as all or part of the Nadia River System or the Kulti Rivers, upon which the life of Calcutta as a city and port depended?

(iii)   Could the attractions of the Ganges-Padma-Madhumati River line displace the strong claims of the heavy concentration of Muslim majorities in the districts of Jessore and Nadia without doing too great a violence to the principle of our terms of reference?

(iv)   Could the district of Khulna usefully be held by a State different from that which held the district of Jessore?

(v)   Was it right to assign to Eastern Bengal the considerable block of non-Muslim majorities in the districts of Malda and Dinajpur?

(vi)   Which State's claim ought to prevail in respect of the Districts of Darjeeling and Jalpaiguri in which the Muslim population amounted to 2.32% of the whole in case of Darjeeling, and to 23.08% of the whole in case of Jalpaiguri, but which constitute an area not in any natural sense contiguous to another non-Muslim area of Bengal?

(vii)   To which State should the Chittagong Hill Tracts be assigned, an area in which the Muslim population was only 3% of the whole, but it was difficult to assign to a State different from that which controlled the district of Chittagong itself?

After discussing, examining and looking at the seven questions, in the final report Radcliffe writes:

> after much discussion, my colleagues found that they were unable to arrive at an agreed view on any of these major issues. There were of course considerable areas of the Province in the south-west and north-east, which provoked no controversy on either side: but, in the absence of any reconciliation on all main questions affecting the drawing of the boundary itself, my colleagues assented to the view at the close of our discussions that I had no alternative but to proceed to give my own decision.[4]

Like in the case of Punjab, in Bengal too irrigation and canal systems were interdependent as a solution, Radcliffe writes:

> I have done what I can in drawing the line to eliminate any avoidable cutting of railway communications and of river systems, which are of

importance to the life of the province: but it is quite impossible to draw a boundary under our terms of reference without causing some interruption of this sort, and I can only express the hope that arrangements can be made and maintained between the two states that will minimize the consequents of this interruption as far as possible.[5]

Under the Bengal Boundary Commission award in the East a significant area of border was demarcated in the middle of the river course, which due to its deltaic nature keeps on shifting its course and creating excessive meandering.[6] Radcliffe did not elucidate the fate of *Chars* which is a common feature of rivers in Bengal. Some *chars* are so small that any flooding leads to their disappearance, but some are very large where a village can be settled. These sand silted *chars* keep on appearing and reappearing and has also became a cause of disputes between India and East Pakistan.[7] To interpret and resolve such leftover disputes, in 1948 India and Pakistan agreed to set up a tribunal under Algot Bagge, former member of the Supreme Court of Sweden. The tribunal's task was to interpret and re-interpret Radcliffe's decisions. The two countries appointed one Judge each to represent them in the tribunal. India appointed Chandrashekhar Iyer while Pakistan was represented by Mohammad. Shahabuddin. The Tribunal submitted its final report on 5 February 1950. Among many of the conflicting interpretations, only four of them came before the tribunal where two were on the western and two on the northeastern part of the boundary of East Pakistan.[8] Three out of four such disputes were due to the flow of border rivers.

The first dispute taken by the tribunal was on fixing the boundary between Murshidabad district of West Bengal and Rajshahi district of East Pakistan including the *thanas* (police stations) of Nawabganj and Shibganj of pre-partition Malda district. They run along the River Ganges between India and East Pakistan. Putting his case Justice Shahbuddin argued for a flexible border in the middle of the river while India stated that

> the district boundary on the date of the Award must be ascertained and demarcated. If this is impossible, the midstream line of the River Ganges and the land boundary will be demarcated within one year from the date of the publication of this Award.[9]

In its award the tribunal stated that:

> if the demarcation of this line is found to be impossible, the boundary between India and Pakistan in this area shall then be a line consisting of the land portion of the above mentioned boundary [in first section of the decision the chairman of the tribunal talked about the boundary as it was in the Radcliffe's report] and of the boundary following the course of the midstream of the main channel of the River Ganges as determined on the date of demarcation and not as it was on the date of

the Award. The demarcation of this line shall be made as soon as possible and at the latest within one year from the date of publication of this decision.[10]

The second dispute arose on the *Mathabhanga* River, a tributary of the Ganges River, and flows between India and East Pakistan (now Bangladesh). The dispute was over a portion of a common boundary which lies between the points on the River Ganges from where the channel of the River Mathabhanga takes off according to Radcliffe's award and the northernmost point where the channel meets the boundary between the *thanas* of Daulatpur and Karimpur, according to that award.[11] The dispute was, mainly, a result of a mapping error committed by the Radcliffe commission due to changing tracks of the rivers in the region. The map showed the existence of the river at a different place from where it was at the time tribunal looked into it. Over the course of time, the Mathabhanga River had slightly shifted to the west direction. To address this India argued for the border to follow the Radcliffe line, whereas Pakistan wanted a flexible border following river even if it had shifted.[12] The Tribunal's award states

> The boundary between India and Pakistan shall run along the middle line of the main channel of the River Mathabhanga which takes off from the River Ganges in or close to the north-western corner of the district of Nadia at a point west – south-west of the police station and the camping ground of the village of Jalangi as they are shown on the air photograph map of 1948, and then flows southwards to the northernmost point of the boundary between the thanas (police station) of Daulatpur and Karimpur.[13]

The second part of the award states that

> The point of the off-take of the River Mathabhanga shall be connected by a straight and shortest line with a point in the mainstream of the main channel of the River Ganges, the said latter point being ascertained as on the date of the Award or if not possible as on the date ascertained shall be the south-eastern most point of the boundary line in Dispute I, this point being a fixed point.[14]

The third dispute was over the course of the River Kusiyara between India and East Pakistan. India accepted Radcliffe's decision over the dispute. The root of the dispute lies in change in name of the river after it crosses the boundary. Justice Shahbuddin argued that

> The boundary in this area shall run along the southern river, i.e the river wrongly described as Sonai in the Award map, from the point where the land boundary running from the south to the north meets the said river,

to the point from where that river takes its water through Noti Khal from the northern river, i.e the river named on the said map as Bogila, and thence along the latter river to the boundary between the districts of Sylhet and Cachar.[15]

The tribunal's decision was that

From the point where the boundary between the thanas of Karimganj and Beani Bazar meets the river described as Sonai River on the map "A" attached to the Award given by Sir Cyril Radcliffe in his Report of August 13th 1947 (Gobindpur) up to the point marked "B" on the said map (Birasri) the red line indicated on the said map is the boundary between India and Pakistan.[16]

The second part of the decision stated

From the point "B" the boundary between India and Pakistan shall turn to the east and follow the river which according to the said map runs to that point from the point "C" marked on the said map on the boundary line between the districts of Sylhet and Cachar.[17]

Although both India and Pakistan agreed that they would accept Bagge's decision, they were reluctant to proceed further where they apparently lost, so that it could take many years to implement the decision. Later, in September 1958 India and Pakistan agreed on a clause between West Bengal and East Pakistan in the areas of Mahananda, Burung, and Karatoa Rivers, that the demarcation would be made in accordance with the latest cadastral survey maps supported by relevant notification and record of rights[18] They also agreed to demarcate the Piyain and Surma river regions with the relevant cadastral survey maps, and "if necessary, record of rights".[19] In the 1958 agreement, India and Pakistan also agreed to divide Berubari Union No. 12 between them. The two sides agreed that the enclaves in Cooch Behar lower down between Beda thana of East Pakistan and Berubari No. 12 would go to Pakistan. In return Pakistan agreed to give up two chitlands of the old Cooch Behar State adjacent to Radcliffe Line to be included in West Bengal. Pakistan also agreed to drop its claim on Bholaganj near border in West Bengal.[20] This was petitioned in the Calcutta High Court first which put a stay. Later, on a parliamentary legality and constitutional basis, it was referred to the Supreme Court under provisions of Article 143 (1) by the then president of India Dr Rajendra Prasad. In 1960 the Government of India brought an amendment to execute the 1958 agreement. Despite the SC's judgement and amendment the 1958 agreement could not be executed, as other appeals were also filed in the SC. One of the last judgements on the issue was on 29 March 1971 on an appeal made by Sudhanshu Majumdar & Ors.[21] By that time East Pakistan reeled into civil war.

Farakka barrage controversy

The dispute over the River Ganges erupted as a result of India's decision to construct a barrage in West Bengal, known as the Farakka Barrage which is about 11 miles from the borders with Bangladesh that was then East Pakistan.[22] An idea to construct a barrage to solve the siltation problem of the River Hooghly and make it navigable for trade first moved by the engineers in nineteenth century when India was under British rule.[23] The British were concerned about the deterioration of the River Hooghly, on which lie the docks, wharves, and jetties of the Port of Calcutta which had been India's busiest port, providing access for international trade primarily to and from Europe. Ships docking in Calcutta had to travel 125 miles up the Hooghly from the sea. Before moving ahead with their project many studies and investigations were carried out between 1853 and 1961.[24]

The first governmental inquiry into the deterioration of the Hooghly was carried out in 1853 by a three member Hooghly Commission. Of the three, two members found it "very difficult to understand how a river like the Hooghly can do otherwise than deteriorate, however gradual or slow that process may be".[25] They recommended that consideration should be given to the establishment of an alternative port. However, the third member of the commission wrote:[26]

I find nothing to lead us to anticipate any future deterioration beyond such as may arise from a temporary shallowing of some of the difficult channels while a change is going on near it.

In 1864, Hugh Leonard, Superintendent Engineer for the Department of Public Works in Bengal, found that evidences of deterioration in the Hooghly was negligible. However, he felt that it was likely to happen slowly. Then,

Figure 3.1 River Hooghly, Kolkata, West Bengal

Source: Photo courtesy of Sudhakar Kumar Rai

L.F. Vernon-Harcourt carried out two surveys for the Commissioners of the Port of Calcutta in 1896 and in 1905.[27] On the basis of his surveys, Vernon-Harcourt concluded that any slow deterioration could be remedied by river "training". By which he meant concentrating the river and tidal currents into a single channel. His second report brought a sharp rebuttal from the Calcutta port officials who at that stage were more concerned to retain confidence in the commercial viability of the port and had no worries about deterioration.[28]

To investigate the deterioration further, in 1916 the government of Bengal established a committee under Sir Charles Stevenson-Moore. It was to advise on improving the flow of the Nadia Rivers – Bhagirathi, Mathabanga, and Jalangi – into the Hooghly. After his study, Stevenson-Moore concluded that:[29]

> The process of deterioration, if indeed any deterioration has taken place, appears to have been very gradual and the fact that even with the evidence available it is difficult to arrive at a definite conclusion proves that the conditions cannot now be very materially worse than in the past.

The Stevenson-Moore Committee's report was reconsidered in 1939 by T.M. Oag, a Deputy River surveyor working for the Commissioners of the Port of Calcutta. Oag advised to set up a means of controlling headwater flow into the Hooghly as a precaution to prevent any deterioration in its navigability.[30] Again, in 1946, A. Webster, Chief Engineer (Special) of the Port Commissioners, also noted that there are no changes in the river below Calcutta, nevertheless he argued that the river was "dependent for its existence" on the freshwater supplied by the Ganges through the Nadia Rivers.[31]

To an extent, all three reports – by the Stevenson-Moore Committee, Oag, and Webster – were convinced that there is a deterioration of the Hooghly and they talk about the needs to stop any further deterioration.[32]

Besides, construction of the Farrakka barrage over which India and Pakistan (later Bangladesh) differed, there is also an issue of diverting Ganges waters which many argue was first proposed by Sir Arthur Cotton, a British Military engineer, in 1853.[33] Cotton found that the inland water way transport system would be more appropriate in India than the railway system. In this context he proposed the construction of a weir across the Ganges. He gave a lecture at Calcutta favouring a dam at Rajmahal and a canal for irrigation, navigation, and supply of water to Calcutta.[34] The canal he proposed would have had a capacity of about 7,500 cusecs, which is one-fifth of the capacity of the canal later built as a part of the Farakka barrage Project.[35]

After 1947, further research on siltation was carried out by using models of the Hooghly and Calcutta port which were constructed for this purpose at the Central Water and Inland Navigation Commission, Research

Station, Poona. They were reviewed in 1952 by a committee of Indian hydraulics experts under the chairmanship of Man Singh which concluded that Hooghly had deteriorated in the stretch between Nabadwip and Calcutta. The committee found that the signs of deterioration were less pronounced lower down the river. To address the situation, it found remedy in building Farakka barrage.[36] In 1960 the sanctioned project was approved by the Ministry of Transport and Shipping. Bihar and Uttar Pradesh were opposed to this project, as it affects water flows and siltation in the two states.[37]

The objective of Farakka was to divert water away from the River Ganges into the Hooghly, thereby flushing it free of silt. Actual work on the barrage started in 1961. At that time, Kapil Bhattacharya, superintending engineer with the Irrigation and Waterways Directorate of the West Bengal, was critical about the project. Bhattacharya believed that Hooghly got silted not because of sedimentation carried over hundreds of kilometres from the Himalayas by the Ganga but because of two dams on the Damodar and Rupnarayan Rivers, western tributaries of the Hooghly.[38] These two dams, he wrote in his report, *Silting of Calcutta Port*, were built without

> taking into consideration flood-tides and tide-borne silts . . . as a result, the Calcutta Port has been killed and the main drainage channel choked, causing repeated flood-havoc on an ever-increasing scale. . . . If my warnings against Farakka Barrage are not heeded, people will have to suffer consequences.[39]

For his warning, dissent, and criticism to Farakka, KL Rao, then Union minister of water resources, called Bhattacharya a traitor. *Anand Bazar Patrika* called him a spy of Pakistan.[40] Consequent upon such attacks Bhattacharya was forced to resign from his office.[41]

The Farakka barrage is about 2,240 metres long. The feeder canal from the barrage, which is about 25 miles long, was completed in 1975 and the barrage came into operation on 21 April 1975.[42] The purpose of barrage is to ensure that the Hooghly River receives, however low the flow of the Ganges may be, up to 40,000 cubic feet per second (cusecs) of water diverted from the Ganges.[43].

During the 1950s and 1960s, Pakistan strongly opposed the construction of the barrage and tried different diplomatic channels to stop its construction. Pakistan argued that the lean flow of the River Ganges of about 50,000 to 55,000 cusecs constituted its normal and basic requirements for irrigation, domestic, municipal, and other uses.[44] Any decrease in such flow of the Ganges would negatively affect irrigation, water supply, fishery production, groundwater tables, and river navigation which is the most common mode of transportation in East Pakistan, and would worsen the problem of salinity.[45] Though talks between India and Pakistan over the Farakka Barrage took place, they could not be concluded. India has maintained that the Ganges is not an "international river". In support of it, India argued that 80% of the Ganges Basin area lies in India.[46]

To resolve this issue, like the Indus Waters Treaty, Pakistan wanted to draw in the international bodies such as the United Nations or the World Bank, but this time India succeeded in keeping away the third party in bilateral problem between the two countries.[47] After its liberation, in 1972, Bangladesh signed the treaty of friendship with India and tried to look out for an amicable solution on water sharing. The joint declaration issued at the end of the visit of the Prime Minister of India, Indira Gandhi, to Bangladesh on 19 March 1972 included the decision to establish a Joint River Commission which would comprise expert members from both countries on a permanent basis. Its work would be to carry out a comprehensive survey of the rivers and river systems shared by the two countries. The Commission also made joint efforts to effectively use the common waters for mutual benefits.[48]

Many in Bangladesh argue that due to Farrakka Barrage, the country has to face multiple problems. During the dry season from January to May every year, drought prevails in Bangladesh because of diversion of some of the waters from the Ganges through the Farakka Barrage and its feeder canal.[49] On the contrary, during the monsoon which lasts from June to September, Bangladesh faces severe flood, when it heavily rains in the region. At that times, it is estimated that at that time, about 2.6 to 3 million hectares in Bangladesh are flooded annually.[50] In an abnormal year, when there is a synchronisation of very heavy rainfall with peak discharges in the Rivers Ganges and Brahmaputra, this figure may reach 6.5 million hectares or some 45% of the total area as happened in 1955 and 1974.[51]

To share water of the River Ganges, in 1975 a partial accord was signed under which India's share during each of the four ten-day periods was far less than the 40,000 cusecs it initially demanded. The total share of Bangladesh ranged between 75–80% of the available water, while India, at the same period, received about 23% of waters.[52] The Accord lasted only for 41 days of the lean season of 1975. It expired on May 31, 1975 and was not renewed or replaced by another agreement. Hence, India started withdrawals to the full capacity of the feeder canal of 40,000 cusecs.[53]

To get more waters, Bangladesh raised the issue at international platforms. One such platform was the seventh Islamic Foreign Ministers' Conference, which was held in Istanbul. Turkey supported Bangladesh and expressed its deep concern over the distribution of the waters of what Bangladesh saw as the "international" River Ganges.[54] A joint communiqué was issued at the end of a four-day meeting (12 May to 15 May 1976) which said that the problem is arising out of India's unilateral withdrawal of Ganges waters only resulted in the aggravation of economic hardship and the retardation of the process of national reconstruction in Bangladesh.[55]

At same time, anti-India sentiments were on rise in Bangladesh over the issue of Farakka. Taking into consideration of such sentiments, on 13 May 1976, then Bangladesh High Commissioner to India, Shamsur Rahman, was informed about the Government of India's regret and concerns at the failure of his government in checking inflammatory propaganda against India over the Farakka barrage foster ill will and hostility towards India.[56] Shamssur Rahman was told that the Bangladesh government has a responsibility to check those engaged in unnecessarily provoking anti-India sentiments in his country.[57] It had little effect on then prevailing situation which was further tensed after prominent Bangladesh leader, Maulana Mohammad Abdul Hamid Khan Bhashani, issued a statement threatening to lead a political march on 16 May to "demolish" Farakka barrage in West Bengal. However, a letter from Indira Gandhi stopped him to move beyond a point.[58]

A few days later, on 21 August 1976, to India's chagrin, Bangladesh took the Ganges water issues to the United Nations. Both sides prepared their own White papers to respond each other in the UN.[59] Before the UN came into the picture, in his statement, Rear Admiral Mosharraf Hossain Khan, Deputy Chief Martial Law Administrator of Bangladesh raised the issue of a large withdrawal of waters by India and stated that the observers from Bangladesh were not allowed to take stock of the amount of waters India withdraws from Farakka.[60] About the Bangladesh decision to raise the issue before the then forthcoming session of the United Nations General Assembly Khan said that the "world opinion would judge it. Bangladesh would get wide support from the fellow members of the United Nations and truth shall prevail".[61] However, Bangladesh was not able to muster enough support in favour of its resolution. Therefore, a Consensus Statement was adopted on 26 November 1976.[62]

In the UN India referred to the Helsinki Rules on water sharing and cited Article IV which allows the basin state to reasonably use transboundary river waters within its territory. Bangladesh's argument was based on principles of reasonable and equitable share from River Ganges.[63] Bangladesh also invoked the theory of injury claiming that the injury caused to Bangladesh through the diversion of the waters to the Hooghly River was clear and substantial, quoting Principle 21 of the Declaration on the Human Environment.[64]

The Consensus Statement adopted by the UN General Assembly on 26 November 1976 proved to be the beginning of the India Bangladesh negotiations over the Ganges water. After negotiations, in 1977 India and Bangladesh signed a water sharing agreement on the River Ganges. This was largely possible as the then newly formed Janata Party government (1977–1980) in India was not shackled by previous positions and was ready to enter into a new era of relationships with Bangladesh. The Janata government also had in mind whatever embarrassment India faced due to the UN Consensus Statement.[65]

The Ganges River water sharing agreement was for five years which expired on 31 May 1982, at the end of the dry season for that year. The mandate for the Commission with regard to the augmentation proposal which ended in 1980 was not renewed.[66] By this time, the Indian National Congress was back in power in India while in Bangladesh Abdus Sattar was the President. Sattar visited India and discussed the issue of water sharing from River Ganges few days before the expiry of the 1977 agreement but rather than agreeing to extend the Agreement the two countries decided to ink a new agreement. A Memorandum of Understanding (MoU) on the sharing of the Ganges was signed in New Delhi on 7 October 1982. In this MoU, except for deleting the guarantee clause, by and large, all provisions were similar to the 1977 water sharing agreement.[67] At that time, it seems, the MoU was the best possible deal that Bangladesh could obtain.[68]

With the end of the dry season of 1984, the 1982 MoU expired. Thereafter, India refused to extend the arrangements under the 1982 MoU for three more years as proposed by Bangladesh.[69] No new agreement was reached during the monsoon season of 1984, and the dry season of 1985 started without any agreement in place.[70]

In 1985, on the sidelines of the Commonwealth Heads of Governments Summit, then President of Bangladesh, General H.M. Earshad and the then Indian Prime Minister Rajiv Gandhi, met at Nassau, in the Bahamas. There the two leaders agreed that India and Bangladesh would sign a MoU, for another three years, reiterating the water sharing formula for each country under the 1982 MoU scheme, and set the terms of reference of a joint study to be undertaken by experts from the two sides on the common water resources.[71] Later, in a meeting of irrigation ministers from the respective countries in New Delhi from 18 to 22 November, a new MoU was agreed.[72] Such ad hoc arrangements continued till 1996 when India and Bangladesh agreed to sign a water sharing treaty.

1996 Ganges water treaty

The 1996 Ganges water sharing treaty was the result of I.K. Gujral's, then Minister for External Affairs of India, policy of extending stretched friendly arm towards the neighbours. This was also supported by the then chief minister of West Bengal, Jyoti Basu. Unlike earlier when the Ganges sharing arrangement was termed as a "Partial Accord", an "Agreement", and a "Memorandum of Understanding", in 1996 it was called "Treaty" which implies a stronger political commitment on the part of the signatories. Second, whereas the 1977 Agreement and the two MoUs of 1982 and 1985 were signed by ministers, either of irrigation or foreign affairs, the Treaty was signed by the two Prime Ministers[73] H.D. Devegowda, then prime minister of India, and Sheikh Hasina, prime minister of Bangladesh. In addition, the Treaty was to remain in force for a period of 30 years, and "shall be renewable on the basis of mutual consent".[74]

The Ganges Water sharing Treaty of 1996 addressed two important concerns – Farakka and the idea of "augmentation". The Indian proposal was for the augmentation of the water-short Ganga from the water-surplus Brahmaputra through a huge link canal from Jogighopa to Farakka, running right across Bangladesh. On the other hand, Bangladesh proposed the augmentation from within the Ganga system by storing its monsoon flows behind seven high dams in Nepal. The two sides had serious reservations on the other's proposal which could not be agreed on despite a series of discussions. This was finally addressed by the 1996 water sharing treaty between the two countries.[75] Yet problems remain, as Punam Pandey points out on three of them. The 1996 treaty faced its first test only a few months after it came into force because the actual availability of the waters of the Ganga at Farakka turned out to be far less than the average flow of Ganga, during the period from 1948 to 1988, as reiterated in the IXth schedule of the treaty – and the flow of water in Padma River was not according to the treaty.[76] The second issue, is related to the discrepancy between the quantum of water released at Farrakka barrage in India and that arriving at the Hardinge bridge – 170 kms downstream – in Bangladesh, which became a major bone of contention between two countries. Third, there is the more complex problem of Gorai hump. Bangladesh's grievance about diversions by India from the Ganga at Farakka has caused acute distress in the South West Khulna region on account of salinity ingress and a shortage of water for agriculture, fisheries, navigation, and sustenance of sundari mangrove species. This area, because of the Gorai spill, which delivers upland forest water supplies to the regions, is left high and dry as Ganga recedes.[77]

At that time, though Jyoti Basu has rejected the theory that the India-Bangladesh agreement on the sharing of the Ganges water would be detrimental to the West Bengal's water interests, he, however, accepted that the treaty was not absolute.[78] This was looked at as detrimental to what was agreed upon in the treaty. In Bangladesh, opposition led by the BNP accused Hasina for agreeing to some "secret clause" and "trade off".[79] To clarify all such rumours and address the opposition parties self-propagated concerns, in an interview with *The Daily Star* on 14 December 1996, I.K. Gujral, said that there is "no secret clause" in the water accord and there has been "no trade off of any kind" between the two countries.[80]

Two decades after the signing of the Ganga Waters Treaty, in May 2017, in a letter to the Indian Prime Minister, Narendra Modi, Mamata Banerjee, the Chief Minister of West Bengal said that the state's experience with the 1996 India-Bangladesh Ganga water sharing treaty was not a happy one.[81] In the letter she pointed out the "adverse" impacts on the availability of water in her state and land erosion in Malda, Murshidabad, and Nadia districts. "The lack of water in the Ganga during the lean season occasionally causes shutting down of the National Thermal Power Plant in Farakka. The promise of making water available has not been fulfilled".[82] She alleged that the Indian government's unfulfilled promises have kept the water problems in West Bengal accruing. "In 2005, the Indian government agreed to take up anti-river-erosion work along a stretch of 120km in the Ganga-Padma River system in Malda and Murshidabad districts, but the commitment was not kept by the Indian government".[83] Due to it, Banerjee's letter said, as per an estimation made in 2015, the damage of public and private properties due to land erosion was INR 707 crore (US $0.935 billion). Since then, further erosion along the river has been noticed and the total loss of private and public properties caused by the Farakka barrage would now exceed Rs 1,000 crore (US $1.321 billion).[84] To overcome the problem of erosion, the West Bengal's Chief Minister requested the Indian Prime Minister

> to direct the Farakka barrage authority to draw up a comprehensive plan in consultation with West Bengal to take up anti-erosion and bank protection work on the entire stretch of the Ganga-Padma River both downstream and upstream.[85]

She maintained that the India-Bangladesh Ganga water sharing treaty of 1996 had not benefited West Bengal or ensured the navigability at Kolkata and Haldia ports.[86] In that letter, Banerjee talked about the flows in the three rivers – Atrai, Tangon, and Punarbhava – which enter into India from Bangladesh before flowing back into that country, have been affected due to construction of barrages and other structures. This affects the water situation in Dakshin Dinajpur district of West Bengal.[87] The letter also informed the Indian prime minister of the "extremely poor" quality of waters, especially of River Churni, an Indo-Bangladesh border river.[88]

Also, over the years, Bhattacharya's foresight about Farakka barrage proves correct and demand for its decommission have been made. On 12 August, 2016, Nitish Kumar, Chief Minister of Bihar formally raised the issue to de-commission Farrakka barrage this time with the Narendra Modi, and again on 20 February 2017. The demand is, mainly because, as has been reported, about 328 million tons of sediment – about 40% of the current sediment volume of the river – is getting trapped behind the Farakka barrage and deposited in the upstream river bed. This leads to severe floods in Bihar.[89] Farrakka barrage has also triggered the funnelling process, due to which upstream riverbank erosion has worsened. It has been reported that more than 3,000 hectares of land has been lost to river erosion in the Murshidabad and Malda districts of West Bengal alone. The extent of erosion in Bihar would be greater.[90]

It is estimated that due to the obstruction of the flow of water by Farakka, each year around 20 billion tonnes of silt has been deposited in the river bed areas of Bihar and Bengal.[91] This has created heavy floods in the region. Even the Central Water Commission of India has accepted that the floods in 2016 in Bihar were due to siltation in Ganges River bed. Not only in Bihar, the barrage has also failed to prove its worth because it has silted the Kolkata port to the extent that it continues to be a sea port.[92]

The Teesta River waters disputes

River Teesta is originates in the Indian state of Sikkim. Its total length is 414 square kilometres out of which 151 kilometres are in Sikkim, 142 kilometres flow along the Sikkim-West Bengal border and through West Bengal, and 121 kilometres in Bangladesh.[93] It is the fourth largest transboundary river between India and Bangladesh. In Bangladesh River Teesta flows through five northern districts of Rangpur Division: Gaibandha, Kurigram, Lalmonirhat, Nilphamari, and Rangpur, comprising a total area of 9,667 square kilometres.[94] Within these five districts Teesta is a catchment area to 35 *upazilas/thanas* (sub-districts/police stations) and 5,427 villages with an estimated population of about 9.15 million people (according to 2011 census).[95] Overall, it is being estimated that about 21 million people in Bangladesh are directly or indirectly dependent on River Teesta for their livelihoods. Percentage wise, River Teesta flood plain covers about 14% of the total cropped area of Bangladesh and provides livelihood opportunities directly to approximately 7.3% of its population.[96] In Bangladesh, water from River Teesta is mainly required during the leanest period from December to March when the water flow temporarily comes down to less than 1,000 cusecs from 5,000 cusecs.[97] The situation turns worse in March and April when there is a decline in groundwater and people do not even get water from tube wells.[98]

Historically, the root of the disputes between India and Bangladesh over sharing of waters from River Teesta lay in the Partition of Bengal. In its report submitted to the BC, Darjeeling and Jalpaiguri districts were demanded by the All India Muslim League on the grounds that they fall in the catchment areas of the Teesta River system. It was thought that by having the two districts, then and in the future, hydro projects over the River Teesta in those regions would serve the interests of the Muslim majority areas of East Bengal.[99] This was opposed by the members of the Indian National Congress and the Hindu Mahasabha. Both of them in their respective reports established India's claim over the two districts. In a final declaration which took into account the demographic composition of the region, administrative considerations, and "other factors", the BC gave a major part of the Teesta's catchment area to India. The main reason to transfer a major part of Darjeeling and Jalpaiguri to India was that both were non-Muslim majority areas. Darjeeling constituted 2.42% of the Muslim population while Jalpaiguri had 23.02% of Muslims.[100] At that time Sikkim was independent and merged with India only in 1975.

After Bangladesh was liberated in 1971, it started discussing water issues from a new angle with India. In 1983, an ad hoc arrangement was made between India and Bangladesh to share waters from Teesta. Under it, 39% was allocated to India, 36% to Bangladesh, and the remaining 25% was left unallocated to be decided later.[101] The Joint Rivers Commission (JRC), as mentioned in the Article IX of the Ganges Water Sharing Treaty of

1996, set up Joint Committee of Experts (JCE) headed by the Secretaries of Water Resources of the Governments of India and Bangladesh to work out arrangements for long term/permanent sharing of the waters of common rivers between the two countries in phases.[102] The Commission accorded priority to the sharing of the Teesta Waters. In 2000 Bangladesh presented the draft of agreement.[103] But it has taken almost ten years to reach at a consensus over the quantity of water sharing between the two countries. The final draft on the Teesta water sharing was accepted by India and Bangladesh in 2010. In September 2011, the then Indian Prime minister Dr Manmohan Singh during his official visit to Bangladesh was all set to sign a treaty on new percentage to share water from River Teesta. Under the 2011 Interim Agreement, India would receive 42.5% and Bangladesh 37.5%[104] of water during the lean season (December to May). The Indian central government accepted the Interim Agreement. However, the West Bengal government did not agree to it. Mamata Banerjee withdrew from the Indian delegation at last minute, about which Rupak Bhattacharjee writes:

> Banerjee subsequently changed her stand, saying she believed that Bangladesh would get 33,000 cubic feet per second (cusec) of water annually, instead of the 25,000 cusecs originally agreed upon. The UPA [United Progressive Alliance] government, however, said that the government of West Bengal was briefed regarding the matter.[105]

Reacting to that political development, Deb Mukherji, India's former High Commissioner to Bangladesh, writes

> Regrettably, the many positive gains from the visit were overshadowed by the last-minute deletion of the sharing of the Teesta waters from the agenda and the related absence of the chief minister of West Bengal from the Indian delegation.[106]

During Narendra Modi's visit to Dhaka in June 2015 to exchange the ratified documents of the Land Boundary Agreement signed between India and Bangladesh in 2011, the Bangladeshi Prime Minister Sheikh Hasina made a request for immediate conclusion of the Interim Agreement on River Teesta.[107] To it, the Indian Prime Minister "conveyed that deliberations are underway involving all stakeholders with regard to conclusion of the Interim Agreements on sharing of waters of Teesta and Feni as soon as possible".[108] At that time Mamata Banerjee also joined Narendra Modi at Dhaka, but remained silent over the issue of Teesta River water sharing.

West Bengal's reluctance to share Teesta waters with Bangladesh is, largely, due to growing water stress in parts of the State. Water from Teesta is important for the irrigation in the five districts of the north Bengal-Coochbehar, Jalpaiguri, South and North Dinajpur, Darjeeling – which constitute some of the poorest blocks in the state. With viable irrigation system these areas have capability to produce three crops in a season.[109] Tea plantation sectors in Darjeeling and Jalpaiguri are recording constant decline in ground water level.[110] Rajib Banerjee, irrigation and agriculture minister of West Bengal, said "We need to irrigate around 1.20 lakh hectares during the lean period, between October and May. The Chief Minister wants to protect the interest of the farmers in the area".[111] To look into the load on River Teesta and other issues, the West Bengal government commissioned a study of the river in 2011 under the hydrologist Kalyan Rudra. He submitted his findings in form of a preliminary report to the West Bengal government in December 2012. The detailed report is not publicly available, however, Rudra's academic writings on Teesta issue are well-known. In one of his papers published in *The Ecologist Asia* in 2003, Kalyan Rudra had been critical to the ongoing projects on the River Teesta, like the Teesta Barrage Project in Jalpaiguri district and the hydropower projects of the National Hydro Power Corporation (stage III and IV) in the Darjeeling district of West Bengal. He is of the view that siltation has been a major problem and projected hydropower capacities of Teesta are decreasing at alarming rates. There is also an increase in evaporation from the reservoirs and seepage of water from canals which has deprived the marginal land of the command area from the water that it was assured during the planning of the project. In some cases, as Rudra finds, dams that were designed to moderate floods have created floods by releasing excess water at the peak of the monsoon.[112] In the paper, Rudra supports an idea to reduce burden on River Teesta by slicing down the number of multipurpose hydro projects on it. This would help it to provide enough waters for irrigation purposes.[113] Physically, it is estimated that in recent years, the water volume in the Teesta River has reached below the 100 cubic metres per second level during the peak summer time, from April and May.[114] To meet some of the river bed related challenges of Teesta, in 2015 the West Bengal government wanted about Rupees 10,000 crore (US $13.28 billion) from the Union government to rejuvenate the riverine area of Teesta.[115]

In April 2017, after a gap of seven years and postponing her visit many times, Sheikh Hasina visited to India. To get a breakthrough over the Teesta and prepare a background for the Hasina's visit in 2017, Bangladeshi officials have made a lot of diplomatic efforts to engage Mamata Banerjee over the Teesta water deal. In June 2016 then Bangladesh High Commissioner to India, Syed Muazzem Ali met Banerjee soon after the assembly election results were declared in West Bengal in which her party – Trinamool Congress – regained power. After coming out from the meeting, Syed Muazzem Ali said "During her last Bangladesh visit, Mamata Banerjee had told us to have confidence in her on the Teesta issue. Today, we have told her that we continue to have confidence in her on the issue".[116] He added that he was hopeful that his country's relation with Bengal would grow simultaneous with the negotiation over the

Teesta and other issues.[117] On the issue of Teesta then External Affairs Minister of India, Sushma Swaraj had said that India was taking initiatives to conclude the deal. She said:

> There are three parties in this. India, Bangladesh and West Bengal government. Assembly elections were held in West Bengal. Now that elections are over and Mamata Banerjee is back as chief minister, the federal Indian government will begin talks with the Mamata government to finalise the Teesta water sharing treaty.[118]

Despite all such engagements, no breakthrough was made, however, Hasina arrived in New Delhi. Before the visit, in an opinion piece for the Indian national daily *The Hindu*, Sheikh Hasina wrote:

> We share our Lalon[119], Rabindranath, Kazi Nazrul, Jibanananda; there is similarity in our language, we are nourished by the waters of the Padma, Brahmaputra, Teesta; and so on. The Sundarbans is our common pride. We don't have any strife over it. Then, why should there be any contention over the waters of common rivers?[120]

Mamata Banerjee reiterated that there is not enough water in the River Teesta to share with Bangladesh. She said, as quoted in Indian media:

> What will I do if there is no water? There is no water in the Teesta. Mukutmanipur has dried up . . . Mahananda has dried up. This is just April. Then May, June are up ahead. By the time the rains start . . . it will be July. So these three months . . . there are water woes.[121]

During their meeting, Narendra Modi, in the presence of Mamata Banerjee, said "I firmly believe that it is only my government and your excellency, Sheikh Hasina, your government that can and will find an early solution to Teesta water sharing".[122] Later, after meeting with Sheikh Hasina, Mamata Banerjee said: "Your problem is water, not Teesta. I am willing to look at any alternate proposal to address your issues. What we can do is that there are many other rivers in the area (India-Bangladesh), we can use water from them".[123]

More than a year after, on 25 and 26 May 2018, Sheikh Hasina visited West Bengal to attend the 49th convocation ceremony of Visva Bharti University in Santiniketan near Kolkata and to receive an honorary Doctor of Literature degree from Kazi Nazrul University. At the Bangladesh Bhavan, in Shantiniketan, Hasina said that "We [India and Bangladesh] have sorted out many bilateral issues. There are still some pending ones but I don't want to spoil the beautiful occasion by referring to them. We want to settle all issues in a friendly ambience".[124] During the visit Hasina met Mamata Banerjee but the two did not develop consensus on the issue.

Regulation of water from River Teesta has also been a cause for disputes within West Bengal which feels if Gorkhaland is formed as a separate state West Bengal will lose its control over the gates of the Gajaldoba dam. In 2011 Mamata Banerjee refused to open gates of the Gajaldoba for Bangladesh, declaring that the water was to be held back for the needs of the population of North Bengal.[125] Gorkhaland, if it comes into existence, includes the Teesta River upstream of the Gajaldoba barrage in the Dooars region in Jalpaiguri district. Water release from Gajaldoba barrage is central to the Teesta River dispute between India and Bangladesh.[126]

There are also other rivers on which India and Bangladesh cooperate. In 2017 India has agreed to provide US $200 million of the fresh Line of Credit for the Buriganga River Restoration Project.[127] At that time, the two Prime Ministers also directed concerned officials to conclude discussions on various aspects relating to sharing of waters of the Feni, Manu, Muhuri, Khowai, Gumti, Dharla, and Dudhkumar Rivers.[128] They also welcomed the visit of an Indian technical team to Bangladesh, establishment of a "Joint Technical Sub Group on Ganges Barrage Project" and study of the riverine border in the upstream area of project.[129]

During Hasina's visit to India on 3–6 October 2019, India and Bangladesh entered into a deal under which India can withdraw about 1.82 cusec of water from River Feni for drinking purpose of the people of Sabroom in the Indian state of Tripura.[130] River Feni has a catchment area of 1.147 square kilometres out of which 535 kilometres are in India. Disputes over sharing its water is going on since 1934. In 2010, Bangladesh allowed India to have a water drinking project and other river protection projects on River Feni.[131]

The water deal over River Feni in 2019 has been criticised by some in Bangladesh and one of them, Abrar Fahad, was a second year student. He was murdered by the members of the Bangladesh Chatra League (BCL). Talking about criticism over the deal and murder of Fahad, Hasina said:

> The volume of water we will give to India is very insignificant and we will use most of the water of this big river. If anyone wants water for drinking purposes, and if we don't give that, it doesn't look good. . . . We always protect our interest in all fields . . . the stretch of the river [Feni] from where India will take water is a catchment area and it belongs to Bangladesh. That is why we signed an MoU with India.[132]

She also recalled the role of the people of Tripura during the Bangladesh's Liberation War of 1971.[133] Targeting the BNP, Hasina said "How could that party [BNP] talk about the issue of Feni River when it forgot to protect the country's interest and realise the rightful demand for water of a big river like the Ganges?"[134]

Another irritant in India-Bangladesh waters disputes is over construction of 162.8 metres, run-of-the river, and multipurpose project – the Tipaimukh dam over River Barak. It has a capacity to produce about 1500 MW of hydroelectricity. There is wide opposition to it in Bangladesh where it is felt that with the help of this project, India will divert water flow from India to Bangladesh.[135] Barrage over Barak River was planned in 1972 and in 1974 the proposed site of the dam was finalised as the Tipaimukh village in Churachandrapur district in the Indian state of Manipur.[136] In 1999, the Tipaimukh hydroelectric project was approved and its construction entrusted to the North Eastern Electric Power Corporation Limited (NEEPCO).[137] This dam was commissioned by India in 2006. In 2009, to resist the dam works, a "Long March" was organized by various Bangladeshi civil society organizations which include Tipaimukh Dam Resistance Committee and Sylhet Division Unnayan Sangram Samiti, supported by leaders of BNP and the Jamaat-e-Islami. It was to go to the Tipaimukh dam site on 10 August 2009 but were stopped short of the international border by the Bangladesh Rifles (BDR).[138] Due to opposition, Manmohan Singh in 2011 assured Bangladesh that India would not do anything to harm Bangladesh's interests.[139] It is estimated that this dam will displace about 60,000 people in the Manipur, including the indigenous Zeliangrong and Hmar communities, and negatively impact 40,000 people in Bangladesh.[140] It is being forecasted that the dam, after coming into operation, will disrupt the seasonal rhythm of the river, agriculture, irrigation, fisheries, drinking water supply, navigation, and ground water levels in the region.[141]

Conclusion

As discussed in this chapter, despite having a good bilateral relationship, water disputes exist between India and Bangladesh. There is an increase in the demands of the waters of the transboundary rivers in both countries. West Bengal does not want to share the newly agreed percentage of Teesta waters which, as it maintains, would affect the demand in the State. In 2019, after Modi returned in power, External Affairs Minister of India, S. Jaishankar, met his Bangladeshi counterpart, AK Abdul Momin, on the sidelines of the CICA (Conference of Interaction and Confidence measures in Asia) in Tajikistan. In the meeting, Momin expressed a hope for an early conclusion of Teesta waters treaty.[142] Under Article 253 of the Indian constitution, the government of India can take unilateral decision on Teesta waters. However, in 2015, then External Affairs Minister of India Sushma Swaraj stated that "Any understanding between governments of India and Bangladesh will not be enough as no decision is possible without consulting the state (West Bengal) government".[143] The Union government of India has maintained that position since 2015.

Besides fulfilling its water demands, Bangladesh also needs hydroelectricity to meet its growing power demands. During the visit of Sheikh Hasina in 2017, India agreed to export more MWs of electricity to Bangladesh which is in addition to the 500 MWs that India earlier exported to it. In 2016, joined with India and Bhutan, Bangladesh has invested US $1 billion in the 1,125MW Dorjilung hydropower project in Lhuentse, Bhutan.[144] Once completed electricity would be imported by Bangladesh from this project. In August 2018 Bangladesh also signed an MoU with Nepal on generation of hydroelectricity.[145]

Notes

1 Ahmad, Qazi Kholiquzzaman. (2005). "India-Bangladesh Co-operation on Transboundary Rivers: Revisiting the Unrealized Opportunities and Unmitigated Challenges". In Haider, Salman (ed) *India-Bangladesh: Strengthening the Relationship*. Chandigarh: Centre for Research in Rural and Industrial Development.
2 "Bangladesh, a country that shares 54 rivers with India" *DNA* 2011, 7 September. https://www.dnaindia.com/india/report-bangladesh-a-country-that-shares-54-rivers-with-india-1584128. Accessed on 12 January 2018.
3 Partition Proceedings Volume VI Commission papers, Reports of the Members and Awards of the Chairman of the Boundary Commissions. (1950). Superintendent, Government Printing West Bengal Government Press, Alipore, p. 117.
4 Ibid.
5 Ibid., pp. 117–118.
6 The Gazzete of India Extraordinary Part-I & – Section 1, Ministry of External affairs, Government of India Retrieved from www.pib.nic.in/archive/docs/DVD_13/. . . BR/EXT-1950-05-02_1259.p; Shewly, H.J. (2008). "Border Management and Post 9/11 Security Concerns: Implications for the India Bangladesh Border". Master's Thesis, p. 35. etheses.dur.ac.uk/223. Access on 24 June 2016.
7 Ibid.
8 Ibid.
9 Ibid.
10 Ibid.
11 Ibid.
12 Ibid
13 The Gazzete of India Extraordinary Part-I – Section 1, Ministry of External affairs, Government of India Retrieved from www.pib.nic.in/archive/docs/DVD_13/. . . BR/EXT-1950-05-02_1259, p. 61.
14 Ibid.
15 Ibid., p. 62.
16 Ibid., p. 63.
17 Ibid.
18 The Acquired Territory (Merger) Act, 1960. Act 64 of 1964 Retrieved from bombayhighcourt.nic.in/libweb/actc/1960.64.pdf. Accessed on 18 January 2017.
19 Ibid.
20 Ministry of External Affairs, Government of India "Agreement Relating to Border Disputes (East Pakistan). https://mea.gov.in/bilateral-documents.htm?dtl/5888/Agreement+relating+to+Border+Disputes+East+Pakistan. Accessed on 11 March 2018.
21 "Union Of India vs Sudhansu Mazumdar & Ors on 29 March, 1971" *Indian Kanoon*. https://indiankanoon.org/doc/490079/. Accessed on 11 May 2017.
22 Rashid, Haroon Noor. (2002). *Indo-Bangladesh Relations: An Insider View*. New Delhi: Har Anand Publications.
23 Iyer, Ramaswamy. (2002). "Three Waters Treaty". In Sahadevan, P. (ed) *Conflicts and Peacemaking in South Asia*. New Delhi: Lancers Publication, pp. 365–395.

24 Crow, Ben, Alan Lindquist and David Wilson. (1995). *Sharing the Ganges: The Politics and Technology of River Development.* New Delhi: Sage.
25 Ibid., p. 26.
26 Ibid.
27 Ibid.
28 Ibid.
29 Cited in Ibid., p. 28.
30 Ibid.
31 Ibid.
32 Ibid.
33 Ibid., p. 32 & Reid, Elizabeth (Cotton) Hope & William Digby (1900) *General Sir Arthur Cotton, R. E., K. C. S. I.: His Life and Work.* Hooder & Stoghtun. https://archive.org/details/generalsirarthu01digbgoog/page/n2/mode/2up. Accessed on 12 November 2019.
34 Ibid.
35 Ibid.
36 Bhasin, Avtar Singh. (2003). *India-Bangladesh Relations Documents -1971–2002. Volume II.* New Delhi: Geeta Press, p. 625.
37 Ibid.
38 Chari, Mridula. (2016, 1 September). "Over 50 Years Ago, Bengal's Chief Engineer Predicted That the Farakka Dam Would Flood Bihar". *Scroll in.* https://scroll.in/article/815066/over-50-years-ago-bengals-chief-engineer-predicted-that-the-farakka-dam-would-flood-bihar. Accessed on 15 March 2019.
39 Ibid.
40 Ibid.
41 Ibid.
42 Salman, M.A. and K. Uprety. (2002). *Conflicts and Cooperation on South Asia's International Rivers: A Legal Perspective.* Washington, DC: The World Bank.
43 Ibid.
44 Ibid.
45 Ibid.
46 Ibid.
47 Verghese, B.G. and Ramaswamy R. Iyer. (1994). *Harnessing the Eastern Himalayan Rivers: Regional Cooperation in South Asia.* New Delhi: South Asia Books.
48 Bhasin, Avtar Singh (ed). (1996). *India-Bangladesh Relations, 1971–1994, Documents.* New Delhi: Siba Exim Pvt. Ltd, p. 88.
49 Ibid.
50 Ibid.
51 Ibid.
52 Ibid.
53 Ibid.
54 Khan, Z.A. (1976). *Basic Documents on Farakka Conspiracy.* Dacca: Khoshroz Kitab Mahal, p. 152.
55 Ibid.
56 Bhasin, Avtar Singh. (2003). *India-Bangladesh Relations Documents -1971–2002. Volume II.* New Delhi: Geeta Press, p. 667.
57 Ibid., pp. 668–669.
58 Ibid.
59 Rashid, Haroon Noor. (2002). *Indo-Bangladesh Relations: An Insider View.* New Delhi: Har Anand Publications, p. 58.
60 Bhasin, Avtar Singh. (2003). *India-Bangladesh Relations Documents -1971–2002. Volume II.* New Delhi: Geeta Press, p. 680.
61 Ibid.
62 Ibid.

63 Salman, M.A. and K. Uprety. (2002). *Conflicts and Cooperation on South Asia's International Rivers: A Legal Perspective*. Washington, DC: The World Bank.
64 Ibid.
65 Ibid.
66 Ibid.
67 Ibid.
68 Salman, M.A. and K. Uprety. (2002). *Conflicts and Cooperation on South Asia's International Rivers: A Legal Perspective*. Washington, DC: The World Bank.
69 Ibid.
70 Ibid.
71 Ibid.
72 Ibid.
73 Ibid.
74 Iyer, Ramaswamy. (2002). "Three Waters Treaty". In Sahadevan, P. (ed) *Conflicts and Peacemaking in South Asia*. New Delhi: Lancers Publication, pp. 365–395.
75 Ibid.
76 Pandey, Poonam. (2012). "Revisiting the Politics of Ganga Water Dispute Between India and Bangladesh". *India Quarterly*, Vol. 68, No. 3, pp. 267–281.
77 Ibid.
78 Bhasin, Avtar Singh. (2003). *India-Bangladesh Relations Documents -1971– 2002. Volume II*. New Delhi: Geeta Press, p. 1108.
79 Ibid.
80 Ibid.
81 "Mamata at It Again". *The Daily Star*, 2017, 26 May. www.thedailystar.net/ frontpage/mamata-hits-back-1410946. Accessed on 26 May 2017.
82 Ibid.
83 Ibid.
84 Ibid.
85 Ibid.
86 Ibid.
87 Ibid.
88 Ibid.
89 Islam, Nazrul. (2017, 13 March). "Farakka Barrage Is Hurting Bangladesh and India". *The Daily Star*. www.thedailystar.net/perspective/farakka-barrage-hurt ing-bangladesh-and-india-1374838. Accessed on 13 March 2017.
90 Ibid.
91 Ibid.
92 Ibid.
93 Political Economy Analysis of Teesta River Basin. (2013). *The Asia Foundation*. https://asiafoundation.org/resources/pdfs/TheAsiaFoundation.PoliticalEcono myAnalysisoftheTeestaRiverBasin.March20131.pdf. Accessed on 3 March 2017.
94 Ibid.
95 Ibid.
96 Ibid.
97 Bhuiyan, Manash Pratim. (2015, 14 June). "Bangladesh Looks to Resolve Teesta Water Disputes with India". *The Quint*. www.livemint.com/Politics/feg4naF tQrlaRtBRjP2jyK/Bangladesh-looks-to-resolve-Teesta-dispute-with-India.html. Accessed on 2 March 2017.
98 Hossain, Md Akhlak. (2017, 25 January). "Teesta Water Pact an Urgent Need for Bangladesh: Experts". *Daily Sun*. www.daily-sun.com/post/201022/Teesta-water-pact-an-urgent-need-for-Bangladesh:-Experts. Accessed on 5 March 2017.

99 Partition Proceedings Volume VI Commission papers, Reports of the Members and Awards of the Chairman of the Boundary Commissions. (1950). Superintendent, Government Printing West Bengal Government Press, Alipore, p. 79.
100 Ibid.
101 Khusru, Syed Munir (2017, 12 April) "A short History of Big Deal" *The Daily Star*. https://www.thedailystar.net/op-ed/politics/short-history-big-deal-1389742. Accessed on 18 April 2018.
102 Salman, M.A. and K. Uprety. (2002). *Conflicts and Cooperation on South Asia's International Rivers: A Legal Perspective*. Washington, DC: The World Bank.
103 Ibid.
104 Khusru, Syed Munir (2017, 12 April) "A short History of Big Deal" *The Daily Star*. https://www.thedailystar.net/op-ed/politics/short-history-big-deal-1389742. Accessed on 18 April 2018.
105 Bhattacharjee, Rupak. (2015). "Why the Teesta Matters". *Himal South Asia*. http://himalmag.com/teesta-matters/. Accessed on 14 March 2017.
106 Mukharji, Deb. (2011, 10 September). "Big Gains in Dhaka, Over to Bengal for Teesta,". Ministry of External Affairs, Government of India. www.mea.gov.in/articles-in-indianmedia.htm?dtl/13640/big+gains+in+dhaka+over+to+bengal+for+teesta. Accessed on 3 March 2017.
107 Ministry of External Affairs, Government of India, Joint Declaration Between Bangladesh and India During Visit of Prime Minister of India to Bangladesh. "Notun Projonmo – Nayi Disha". www.mea.gov.in/bilateral-documents.htm?dtl/25346/joint+declaration+between+bangladesh+and+india+during+visit+of+prime+minister+of+india+to+bangladesh+quot+notun+projonmo++nayi+dishaquot. Accessed on 15 June 2017.
108 Ibid.
109 Das, Madhuparna. (2015, 13 June). "Teesta Accord: West Bengal CM Mamta Banerjee May Be Eyeing Bigger Compensation". *The Economic Times*. http://economictimes.indiatimes.com/news/politics-and-nation/teesta-accord-west-bengal-cm-mamata- Accessed on 15 June 2017.
110 Ibid.
111 Ibid.
112 Rudra, Kalyan. (2003, January–March). "Taming the Teesta". *The Ecologist Asia*, Vol. 11, No. 1. www.actsikkim.com/docs/Rudra_Taming_the_Teesta.pdf. Accessed on 8 March 2017.
113 Ibid.
114 "Teesta Has One-Sixteenth of Water Needed". thethirdpole.net, 2017, 4 April. https://www.thethirdpole.net/ en/2017/04/14/teesta-has-one-sixteenth-of-water-needed/. Accessed on 17 April 2017.
115 Das, Madhuparna. (2015, 13 June). "Teesta Accord: West Bengal CM Mamta Banerjee May Be Eyeing Bigger Compensation". *The Economic Times*. http://economictimes.indiatimes.com/news/politics-and-nation/teesta-accord-west-bengal-cm-mamata. Accessed on 15 June 2017.
116 Ghoshal, Annirudh. (2016, 21 June). "Teesta Water Agreement: Hasina Set to Visit India This Year, Meet Mamata Banerjee". *The Indian Express*. http://indianexpress.com/article/india/india-news-india/teesta-water-sharing-deal-agreement-mamata-banerjee-sheikh-hasina-meet-2865445/. Accessed on 8 March 2017.
117 Ibid.
118 Ibid.
119 Lalon Fakir was a mystic, songwriter, social reformer and thinker. Many of his songs are still being sung by folk singers and liked by many in both West Bengal and Bangladesh.

120 Hasina, Sheikh. (2017, 7 April). "Friendship Is a Flowing River: Sheikh Hasina Writes for the Hindu". *The Hindu*. www.thehindu.com/opinion/lead/friendship-is-a-flowing-river/article17854490.ece?homepage=true. Accessed on 7 April 2017.

121 "No Water in Teesta: Mamata". *Business Standard*, 2017, 5 April. www.business-standard.com/article/news-ians/no-water-in-teesta-mamata-117040500898_1.html. Accessed on 12 April 2017.

122 "India, Bangladesh Signed 22 Pacts in Key Sectors, Teesta Issue Unresolved". *The Indian Express*, 2017, 8 April. http://indianexpress.com/article/india/nar endra-modi-sheikh-hasina-india-bangladesh-key-pacts-credit-line-teesta-issue-road-rail-lines-4604774/. Accessed on 8 April 2017.

123 "Mamata Calls Teesta WBs Lifeline". *Pratham Alo*, 2017, 9 April. http://en.prothom-alo.com/bangladesh/news/144797/Mamata-calls-Teesta-WB-s-lifeline-says-it-can%E2%80%99t. Accessed on 9 April 2017.

124 Singh, Shiv Sahay and Subhojit Bagchi. (2018, 26 May). "Want to Settle All Issues in a Friendly Ambience: Hasina". *The Hindu*. http://www.thehindu.com/news/national/want-to-settle-all-issues-in-a-friendly-ambience-hasina/arti cle23994739.ece?homepage=true. Accessed on 26 May 2018.

125 Oak, Gauri Noolkar. (2017, 31 July). "Gorkhaland: Troubles on the Teesta". *The Diplomat*. http://thediplomat.com/2017/08/gorkhaland-troubles-on-the-teesta/. Accessed on 1 August 2017.

126 Ibid.

127 Ministry of External Affairs, Government of India. (2017). "India-Bangladesh Joint Statement". www.mea.gov.in/bilateral-documents.htm?dtl/28362/India__Bangladesh_Joint_Statement_during_the_State_Visit_of_Prime_Minister_of_Bangladesh_to_India_April_8_2017. Accessed on 18 July 2018.

128 Ibid.

129 Ibid.

130 Ibid.

131 "Bangladesh, a country that shares 54 rivers with India" *DNA* 2011, 7 September. https://www.dnaindia.com/india/report-bangladesh-a-country-that-shares-54-rivers-with-india-1584128. Accessed on 12 January 2018.

132 The Daily Star. (2019, 10 October). "Search All University Dorms". https://www.thedailystar.net/frontpage/bangladesh-pm-sheikh-hasina-says-search-all-university-dorms-1811743. Accessed on 10 October 2019.

133 Ibid.

134 Ibid.

135 Habib, Haroon. (2011, November 21). "In Bangladesh Tipaimukh Dam Pact Sparks Fresh Row". *The Hindu*. https://www.thehindu.com/news/international/In-Bangladesh-Tipaimukh-dam-pact-sparks-fresh-row/article13492517.ece.

136 Thakur, Jaya. (2020). "India-Bangladesh Trans-Boundary River Management: Understanding the Tipaimukh Dam Controversy". *ORF*. Issue Brief No. 334. https://www.orfonline.org/research/india-bangladesh-trans-boundary-river-management-understanding-the-tipaimukh-dam-controversy-60419/. Accessed on 28 March 2020.

137 Ibid.

138 Rahman, Mirza Zulfiqar. (2009). "India, Bangladesh and Tipaimukh Dam". *IPCS*. http://www.ipcs.org/focusthemsel.php?articleNo=2945. Accessed on 29 March 2017.

139 Habib, Haroon. (2011, November 21). "In Bangladesh Tipaimukh Dam Pact Sparks Fresh Row". *The Hindu*. https://www.thehindu.com/news/international/In-Bangladesh-Tipaimukh-dam-pact-sparks-fresh-row/article13492517.ece.

140 "Tipaimukh High Dam". *International Rivers*. https://www.internationalrivers.org/resources/tipaimukh-high-dam-3499. Accessed on 21 February 2019.

141 Ibid.
142 Noolkar-Oak, Gauri. (2019, 24 July). "Signing a Meaningful Teesta Treaty". *The Daily Star*. www.thedailystar.net/opinion/news/signing-meaningful-teesta-treaty-1775887. Accessed on 25 July 2019.
143 "Proactive PM: Teesta Deal, Talks With Pakistan, Modi's Israel Tour: 4 Things Sushma Swaraj Told Us". *Hindustan Times*, 2015, 1 June. https://www.hindustantimes.com/india/proactive-pm-teesta-deal-talks-with-pakistan-modi-s-israel-tour-4-things-sushma-swaraj-told-us/story-aAPi8mP8EleG2lBbSooPII.html. Accessed on 15 June 2017.
144 *Bangladesh Recently Announced USD 1B for the Project Druk Green*. www.drukgreen.bt/images/yootheme/NewsEvents/External_News/Bangladesh_to_invest_in_Bhutans_hydropower.pdf. Accessed on 12 July 2019.
145 Ministry of Foreign Affairs, Government of Nepal. "Memorandum of Understanding Between Government of Nepal and Government of Bangladesh". www.moen.gov.np/pdf_files/MoU-between-Nepal-and-Bangladesh.pdf. Accessed on 15 June 2019.

4 River water and Hydroelectric Power Projects issues between India and Nepal

Nepal, like Bangladesh and Pakistan, was not a part of British India. How-ever, since 1816, the British had close relations with the Nepali kings. Other than India, Nepal is the only country where more than 80% of the popu-lation is Hindu. This demographic similarity makes India and Nepal, at people-to-people, level close friends, but political differences between the governments remain. Nepal depends a lot on India because of its geographi-cal location and topography. As Nepal is a landlocked country, India pro-vides transit routes and sea ports facilities to facilitate third country trade to Nepal. India is Nepal's largest trade partner and the largest source of foreign investments.[1] Physically, India and Nepal share about 1880 square kilome-tres of a border out of which 640 square kilometres are a river border. The two countries have 60 boundary- river and rivulets along their border. Of the total length of the Riverine border, 200 kilometres are along the River Mahakali, about 80 kilometres are along the River Mechi River, 20 kilome-tres along the River Gandak, and 57 other river and rivulets amount to 340 kilometres.[2]

Nepal is small but rich with water resources. Political and cultural impor-tance of rivers in uniting, integrating, and defining Nepal is such that like India, rivers' names have been included in the country's national anthem[3]:

Woven from hundreds of flowers, we are one garland that's Nepali Spread sovereign from Mechi to Mahakali.

A playground for nature's wealth unending Out of the sacrifice of our braves, a nation free and unyielding.

A land of knowledge, of peace, the plains, hills and mountains tall Indi-visible, this beloved land of ours, our motherland Nepal.

Of many races, languages, religions, and cultures of incredible sprawl.

This progressive nation of ours, all hail Nepal.

It is estimated that water resources in Nepal can be harnessed to produce more than 42,000 megawatts (MW). However, theoretical capacity (ideal capacity) of Nepal is estimated to be around 83,000 MW, which is more than the combined total of Hydroelectric Power (HEP) produced by USA, Canada, and Mexico.[4] Most of the active HEP projects in Nepal are run-of-the river so only rarely can impound year-round reservoir storage with

HEP capacity.[5] As a result, a gap remains and also year-round production of hydropower from all plants are almost impossible. Basin wise potential of HEP generation in Nepal is:

Table 4.1 Electricity Demand Forecast Report (2015–2040) in Nepal

Major river basins	Theoretical power potential in MW		Total	Technical potential		Economic potential	
	Major river Courses having catchment areas above 1000 square kilometres	Small rivers Course having Catchment areas 300 to 1000 square kilometres		Number of project sites	Technical potential in MW	Number of project sites	Economic potential in MW
Sapta Kosi	18750	3600	22350	53	11400	40	10860
Sapta Gandaki	17950	2700	20650	18	6660	12	5270
Karneli and Mahakali	32680	3500	36180	34	26570	9	25125
Southern river	3070	1040	4110	9	980	5	878
Country total	72450	10840	83290	114	45610	66	42133

Source: Electricity Demand Forecast Report (2015–2040) Government of Nepal, Water and Energy Commission Secretariat, https://www.wecs.gov.np/uploaded/Electricity-Demand-Forecast-Report-2014-2040.pdf, P 8 Accessed on 21 January 2019

Nepal is an upper riparian to almost all transboundary rivers flowing between India and Nepal. Water resources have been getting due prominence in the agenda of their bilateral cooperation, for a long time.[6] With a view to optimising the benefits and addressing the water related problems between India and Nepal, the two countries have set up three-tier mechanisms called the Joint Ministerial Commission for Water Resources (JMCWR), Joint Committee on Water Resources (JCWR), and Joint Standing Technical Committee (JSTC) to implement agreements and treaties and also address problems of flood and inundation.[7] There is also a Joint Committee on Inundation and Flood Management (JCIFM) which explicitly deals with the issues of inundation, embankments, and flood forecasting.[8]

As Nepal does not have adequate technologies, to explore and use its vast HEP potential, since the British colonial days, it has entered into a number of agreements and treaties with India. Most of the India-Nepal water related projects were started when monarchy was strong in Nepal. After democracy was restored in 1990, the speed of development of HEP project construction was significantly hampered. Even the then ongoing Maoist movement affected the pace of some of the projects. In addition, there were also a rise of environmental protest movements in Nepal.[9] However, as the government

needed investments they opened the Hydroelectric sector of Nepal for private investors and for foreign investment companies.[10]

In 2014 India and Nepal signed a Power Trade Agreement which opened
the door for the Nepali power traders to export electricity to India.[11] It is
maintained that significant electricity exports to India from Nepal will begin
only from the year 2025, after more number of transmission lines will be
constructed. Afterwards, Nepal is estimated to import around 0.7 billion
kiloWatt Hours (bkWH) in 2020 but by 2025 it would export around 18
bkWH. This will increase to 65 bkWH by 2030 and to 113 bkWH by 2040.
This trade would help Nepal to accrue around 310 billion Nepali Rupees
(NPR) (US $2.56 billion) in 2030, 840 billion NPR (US $6.90 billion) in
2040, and 1,069 NPR (US $8.84 billion) in 2045, at 2011–12 prices.[12]
Despite the significance of India in the Nepal's hydropower sector, a number
of people in Nepal do have reservations against the Indian projects. Some
of these HEP projects also face protests in India, particularly from people
who live in catchment areas of India-Nepal rivers. This chapter looks at
India-Nepal HEP projects and examines concerns expressed by the Nepali
about them.

Beginning of India-Nepal cooperation on hydroelectricity projects

In Nepal a first move to develop a large canal irrigation system was done
during the Rana's rule (1846-1951). The first large-scale State supported
irrigation canal (the Chandra Nahar) was constructed in 1922. It has a net
command area of about 10,000 hectares.[13] The British engaged in water
negotiations with Nepal's government to irrigate lands in the United Province. The first such recorded water resource negotiations between Nepal
and the British India occurred between 1910 and 1920 when British India
needed to harness the Sarda or Sharda (also known as Kali in upper stream
then Mahakali) River, which formed the north-western boundary between
Nepal and British India, to develop irrigation in the United Province (Uttar
Pradesh). Nepal agreed to the 1920 Sarda treaty, involving an exchange of
territory. The British India government formalised with its Nepali counterpart in 1920, the negotiations of the Sarda Treaty in the form of an Exchange
of Letters. These letters gave permission to Colonel Kennion to send his staff
to build barrage on the Mahakali River at Banbassa (now in Uttrakhand)
bordering the present Mahendra Nagar in Nepal.[14] The main features of
Sarda project assessment treaty were (written verbatim)[15]:

(a) The Nepal Government will have a right to a supply of 460 cusecs of
    water and, provided the surplus is available, for a supply of up to 1,000
    cusecs when cultivation grows at any future time from the Sarda canal
    head work during the Kharif i.e., from 15 May to 15 October; and of

150, cusecs during Rabi i.e., from the 15 October to 15 may, the canal head being in the latter period alternately closed and opened for ten days at a time running 300 cusecs whenever the canal is open.

(b) There is order to give those supplies all necessary works such as the canal head with regulating gates, quarters from the canal staff, i.e., on the left bank of the river and also under-sluices for the purpose of maintaining an open 298 channel from the river to the canal head will be done by the Government of India at their own expense on the understanding that they shall retain full and entire control of the work with this undertaking and that they shall supply to Nepal the quantity of water agreed to free of any charge.

(c) That the Nepal Government would transfer necessary land for the construction and maintenance of canal works which is provisionally estimated at 4,000 acres and would receive land equal in area from the British Government. The land to be taken from Nepali territory will, after demarcation, be measured and then land equal in area to it will be given to Nepal by the said Government.

In 1947, India became independent and successor State to the British India. In 1950, to improve their bilateral relationship, India and Nepal signed a peace and friendship treaty. The 1950 treaty also includes a letter exchanged between the two countries along with that treaty. Para 4 of the letter exchanged along with the India-Nepal friendship treaty stipulated that:[16]

If the Government of Nepal should decide to seek foreign assistance in regard to the development of national resource, or any industrial projects in Nepal, the Government of Nepal shall give first preference to the Government or nationals of India, as the case may be, provided that the terms offered by the Government of India or Indian nationals are not favourable to Nepal than the terms offered by any other Government or by other foreign nationals.

Disputes between India and Nepal over sharing the benefits of the Mahakali River started shortly after India's unilateral decision to construct the Tanakpur barrage on India-Nepal border in Uttrakhand in 1983.[17] The latter was 18 kilometres upstream from the Sarda barrage. Nepal had objections with the Sarda treaty and it constantly tried to renew this treaty but it could not and the treaty continued for 76 years, from 1920 to 1996, when it was replaced by the Integrated Development of Mahakali River Treaty.[18] It included development of the Sarda barrage, Pancheshwar project, and Tanakpur barrage.

After negotiations, Tankapur Agreement was signed on 6 December 1991. This Agreement provided for the construction of the left afflux bund (the retaining wall) on Nepali territory for which the Nepal provided

2.9 hectares of land to India.[19] This agreement is being considered as a hasty decision and a lopsided one in favour of India, so it was highly criticised. People in Nepal accused the then Nepali Government, led by Girija Prasad Koirala, for not appreciating the legal, socio-economic, and political ramifications involved in the issue, or for deciding to overlook them to appease India.[20]

The issue raised in the objections dealt primarily with a concern for Nepal's territorial sovereignty and a belief that Nepal had not benefited from the Project as much as India had. Those opposing the agreement argued that because the agreement dealt with natural resources it fell under the articles of the constitution and required ratification by a two-thirds majority of Parliament. A writ petition was filed in the Supreme Court, with the Prime Minister as one of the respondents, challenging the validity of the Tanakpur Agreement. The Supreme Court gave its verdict in December 1992 and concluded that the Tanakpur Agreement was, indeed, a Treaty that required ratification by the Parliament and was not a mere Memorandum of Understanding (MoU).[21] Under the treaty, Nepal's rights over Mahakali have been limited to as low as 4%. To hide their failure the political parties passed a stricture on the treaty through *sankalpa prastav* in parliament.[22] The constitutional committee incorporated a provision in the constitution that any agreement covering utilisation and distribution of Nepal's natural resources would need to be approved by a two-thirds majority of the members of both the houses present and voting in a joint session of Parliament.[23]

In 1996 Mahakali treaty was signed between India and Nepal, Tanakpur project was a part of it. Article 2 of The Mahakali treaty states that:[24]

1   For the construction of the eastern afflux bund of the Tanakpur Barrage, at Jimuwa and tying it up to the high ground in the Nepali territory at EL 250 M, Nepal gives its consent to use a piece of land of about 577 metres in length (an area of about 2.9 hectares) of the Nepali territory at the Jimuwa Village in Mahendranagar Municipal area and a certain portion of the No-Man's land on either side of the border. The Nepali land consented to be so used and the land lying on the west of the said land (about 9 hectares) up to the Nepal-India border which forms a part of the pondage area, including the natural resources endowment I in within that area, remains under the continued sovereignty and control of Nepal and Nepal is free to exercise all attendant rights thereto.

In lieu of the eastern afflux bund of the Tanakpur Barrage, at Jimuwa this constructed, Nepal shall have the right to:

a)   A supply of 28.35 m³/s (1000 cusecs) of water in the wet season (i.e. from 15th May to 15th October) and 8.50 m³/s (300 cusecs) in the dry season (i.e. from 16th October to 14th May) from the date of the entry into force of this Treaty. For this purpose and for the purposes of Article I herein, India shall construct the head regulator(s) near the left under sluice of the Tanakpur Barrage and also the waterways of the required capacity up to the Nepal-India border. Such head regulator(s) and waterways shall be operated jointly.

b) A supply of 70 million kilowatt-hour (unit) of energy on a continuous basis annually, free of cost, from the date of the entry into force of this Treaty. For this purpose, India shall construct a 132 kilo Volts transmission line up to the Nepal-India border from the Tanakpur Power Station (which has, at present, an installed capacity of 120,000 kilowatt generating 448.4 million kilowatt-hour of energy annually on 90% dependable year flow).

2 Following arrangements shall be made at the Tanakpur Barrage at the time of development of any storage project(s) including Pancheshwar Multipurpose Project upstream of the Tanakpur Barrage:

a) Additional head regulator and the necessary waterways, as required, up to the Nepal-India border shall be constructed to supply additional water to Nepal. Such head regulator and waterways shall be operated jointly.

b) Nepal shall have additional energy equal to half of the incremental energy generated from the Tanakpur Power Station, on a continuous basis from the date of augmentation of the flow of the Mahakali River and shall bear half of the additional operation cost and, if required, half of the additional capital cost at the Tanakpur Power Station for the generation of such incremental energy.

Further, Article 3 of the Mahakali Treaty says that[25]

Pancheshwar Multipurpose Project is to be constructed on a stretch of the Mahakali River where it forms the boundary between the two countries and hence both the Parties agree that they have equal entitlement in the utilization of the waters of the Mahakali River without prejudice to their respective existing consumptive uses of the waters of the Mahakali River. Therefore, both the parties agree to implement the Project in the Mahakali River in accordance with the Detailed Project Report (DPR) being jointly prepared by them. The Project shall be designed and implemented on the basis of the following principles:

a The project shall, as would be agreed between the Parties, be designed to produce the maximum total net benefit. All benefits accruing to both the Parties with the development of the Project in the forms of power, irrigation, flood, control etc., shall be assessed.

b The project shall be implemented or caused to be implemented as on integrated project including power station of equal capacity on each side of the Mahakali River. The two power stations shall be operated in an integrated manner and the total energy generated shall be shared equally between the Parties.

c   The cost of the Project shall be borne by the Parties in proportion to the benefits accruing to them. Both the Parties shall jointly endeavour to mobilize the finance required for the implementation of the project.

d   A portion of Nepal's share of energy shall be sold to India. The quantum of such energy and its price shall be mutually agreed upon between the Parties.

Under Article 10 of the Mahakali treaty "Both the Parties may form project specific joint entity/ies for the development, execution and operation of new projects including Pancheshwar Multipurpose project in the Mahakali River for their mutual benefit".[26] Pursuant to this article of the Mahakali treaty the two governments in 2009 agreed and framed draft of Terms of Reference for setting up the Pancheshwar Development Authority (PDA), as an independent autonomous body, for development, execution, and operation of the Pancheshwar Multipurpose Project as well as finalise its Detailed Project Report (DPR).[27] Proposed dam site is around 2.5 kilometres downstream of the confluence of River Sarju with River Mahakali.[28]

> The project would comprise of a rock-fill dam with central clay core of 311 m height from the deepest foundation level. Two underground power houses at Pancheshwar dam, one on each bank of Mahakali River, each with a capacity of (6x400 MW) with the total installed capacity of nearly 4800 MW are proposed to be constructed. The power plant at main dam will be operated as the peaking station to meet energy demand in India and Nepal.[29]

This project aims at producing hydroelectricity and enhance the food grains production in India and Nepal by providing additional irrigation resulting from the augmentation of dry season flows.[30] Year-round irrigation will be possible in agricultural land in the Kanchanpur district in Nepal due to the enhancement in flows during non-monsoon months. The project is expected to generate 10,055 GWh of energy annually at the Pancheshwar and Rupaligad dam power houses during 90% dependable years.[31] Irrigation benefits in form of annual irrigation will be about 0.43 Mha, out of this, annual irrigation in Nepal would be 0.17 Mha and remaining 0.26 Mha in India. In addition, due to moderation of flood peak at reservoir(s), incidental flood control benefits for both the countries are also envisaged from the project.[32]

During the 3rd meeting of Joint Committee on Water Resources (JCWR) held in September–October 2008 at Kathmandu, it was decided to set up PDA.[33] However, it took four years and only in February 2012 PDA was finally set up. The two sides agreed to fast track completion of the DPR of Sapta Kosi High dam and the Sun Kosi Storage-cum-Diversion scheme by February 2013.[34] Final draft of the DPR was submitted in 2016 and in 2017 revised draft of the DPR was submitted to the project development authority.[35] Water and Power Consultancy Services (WAPCOS) limited, a public undertaking of the government of India, has been engaged in preparing DPR.

Pancheshwar dam project is facing protests both in the Indian state of Uttrakhand and in Nepal. In 1996 after the Mahakali treaty was signed, a section of Nepalis raised various objections over it. A faction of the Maoist party of Nepal led by Krishna Vaidya has demanded for its scrapping.[36] In Nepal some of the analysts continued to be apprehensive and raised the issue of soft loan of $1 billion from India to Nepal. They argue that the project has a "ripple effect" and led the Nepal government to cave in.[37] Some of the clauses in the Mahakali treaty, such as on water allocation and power purchase, and also a strategic issue vis-a-vis the area around the origin of the Mahakali River are still disputed.[38] The two countries have not yet reached at the agreement on the rate at which Nepal will get power. Environmental Impact Analysis (EIA) carried out by WAPCOS in India, as analysed by some scholars, lacks complete analysis of the geological impact of the project. Compared to it, they find, Nepal's EIA of its side is better.[39]

It is apprehended that the border villages in India falling in Mahakali's catchment region are also going to suffer from this project. The 33,108 crores (US $43. 4 billon) Pancheshwar project is estimated to submerge 134 villages in Pittorgarrh and Champawat districts in Uttarakhand in India. In 2017 people from those villages fearing submergence of their villages burnt the copy of DPR and protested against it. Of 134, 123 villages would be rehabilitated still protests are there.[40] The project affectees have been also promised compensation, which many find just a peanut.[41] It is estimated that after coming into operation Pancheshwar dam will affect about 50,000 or more people directly in Pittoragarh, Champawat, and Almora districts of Uttarakhand. In Nepal, it will directly impact near about 3,000 families and about 13,700 hectares of agricultural land.[42] Given its negative impact on the people and complexities of treaty, Ramaswamy R. Iyer, former secretary water ministry, Government of India, was of the view to scrap the Pancheshwar project.[43]

Other major hydroelectricity project India is developing is Arun III in Sankhuwasabha District of Eastern Nepal. It was bagged by Sutlej Jal Vikas Nigam Limited (SJVNL) through International Competitive Bidding. An MoU was signed between Nepal government and SJVNL in March 2008 for execution on Build Own Operate and Transfer (BOOT) basis for a period of 30 years, including 5 years of construction period.[44] In 2014, during the Indian Prime Minister Narendra Modi's visit to Nepal, the two countries signed Project Development Agreement on Arun III. At that time, an important Power Trade Agreement was also signed between the two countries.

It paved the way for the power developers of the two countries to trade electricity across the border without restrictions. Under it, private/public power developers from India have reached agreements with Nepal's Investment Board to develop two mega hydropower projects – Upper Karnali and Arun III.[45]

In May 2018, during Modi's visit to Nepal, foundation stone of 900 MW run-of-the River Arun-III HEP project in Nepal was laid. While laying the foundation stone, Modi and the Nepali Prime Minister K.P. Oli "expressed hope that operationalization of the project would help enhance cooperation in the generation and trade of power between the two countries".[46] They agreed "to enhance bilateral cooperation in power sector in line with the bilateral Power Trade Agreement".[47] The Project Development Agreement for Arun III signed in November 2014 provides 21.9% free power to Nepal for the entire concession period of 25 years.[48] In 2017 the Indian Cabinet had approved the investment proposal for generation component of Arun III at an estimated cost of INR 5,723.72 crore (US $7.57 billion) at May 2015 price level. Later in 2019, it approved INR 1,236.13 crore (US $1.63 billion) at June 2017 price level.[49] The proposed HEP agreement on Arun was critically analysed and questioned by civil society and water experts.[50]

Due to protests and criticisms, India aided projects in Nepal take a long time to come into operation. For example, the feasibility survey of Karnali River project began in 1961. Two surveys were carried out. One was by Nippone Koie Company of Japan and the other was by Snoway Mountain Hydroelectric Authority of Australia. Afterwards, a survey was entrusted to United Nations Development Programme to select project sites as both the reports had given different suggestions.[51] The estimated cost of project was about US $3.2 billion at the 1982 price. After a few hesitations, India agreed to finance the project. Nepal's worry was about the purchase of surplus power which was addressed by India. In 1981, India reiterated its willingness to purchase surplus energy from Karnali hydropower project. Then, in 1983, India and Nepal agreed to execute three multipurpose projects on Karnali River: Karnali, Pancheshwar, and West Rapti.[52]

In 2008, the 900 MW Upper Karnali HEP project was bagged by India's infrastructure conglomerate GMR Group. The project is located at Surkhet, Dailekh, and Accham districts.[53] It is expected to start commercial operation by September 2021.[54] However, in 2013 due to protests the Supreme Court of Nepal put a temporary stay on the construction which was lifted after a month.[55] During the 2016 Madhesi stir against some constitutional provisions in Nepal's new constitution, a petrol bomb was hurled at the GMR office in Kathmandu.[56] In February 2020, the Bangladesh Power Development Board issued a letter of intent to enter into a power purchase agreement for the purchase of the electricity at the rate of 7.712 cents per unit for a period of 25 years.[57]

India also helped Nepal to tap the Trisuli River for hydropower. The idea of tapping the river is as old as 1930 when a scheme to generate 1000 KW energy was envisaged. On Trishali, the agreement between India and Nepal was signed in 1958. The project was estimated to generate 21,000 KW of electricity and cost around 3.5 crores (US $397,444). It was completed in 1966 and the final phase was completed in 1971.[58] Other important hydropower project undertaken by India was the Devighat project in 1976 at a cost of 30 crores. Its capacity was 10,400 KW and was implemented by National Hydro Power Corporation of India. Subsequently, India also aided to complete a few other projects such as Surajpur project completed in 1981, Kalapriya in 1977, and Phursa Khola in 1983.[59]

India is also trying to develop river-linking projects which would link Nepal's Rivers with India. A big move in this direction came in February 2016 when India and Nepal had set up an authority and started working on the first trans-country river-linking project to channel water from Sarda River in Nepal to Sabarmati in the Indian state of Gujarat. As planned, in the first phase, India will develop five reservoirs on the Sarda River.[60] Under the arrangement the dams will be built by notified PDA. At that time "The authority has started working on building five reservoirs on Sarda River", said Sriram Vedire, advisor to the then water resources minister Uma Bharati.[61] The Central Water Commission is working on studying the feasibility of taking Sarda water to Sabarmati via Yamuna and Sukli Rivers. As per the initial plan, excess water stored in reservoirs to be built on Sarda River will be channelled into Yamuna in Uttarakhand, from there to Sukli River in Rajasthan via Haryana, and finally into Sabarmati, about 2,000 kilometres from the main source. The cost of the project is estimated to be over Rs 1 lakh crore (US $132.3 billion) over the next 15–20 years.[62] The Sarda-Sabarmati River-linking project was revived during Modi's Nepal visit in August 2014 – 12 years after it was envisaged by the Atal Bihari Vajpayee-led National Democratic Alliance government in 2002, as part of its river inter-linking project. Modi signed letters of exchange regarding terms of reference of PDA, which will execute the project on river Sarda.[63]

India and Nepal are also working on developing inland waterways for trade and transit. In February 2018, during K.P. Sharma Oli's visit to New Delhi, India and Nepal "recognised the untapped potential of inland waterways to contribute towards overall economic development of the region".[64] The two countries decided to develop inland waterways for the movement of cargo, within the framework of trade and transit arrangements, providing additional access to sea for Nepal.[65]

Kosi River water treaty

The Kosi is regarded as the wildest river with the most devastating effects in the Indian state of Bihar, for which it is also referred to as the "sorrow of Bihar". Because of the seasonal damage it caused, a scheme to attenuate the effects of the Kosi was deemed necessary since the British colonial days. Between the mid-18th century and the mid-20th century, Kosi has changed its track by moving westwards over 100 kilometres during the process it abandoned old channels and formed new ones.[66] The river carried huge sediments which complicated British flood-control efforts. Due to large amount of sediments, whenever attempts were made to build embankments along small sections of the river, the river bed began to rise faster than before due to the deposited alluvium had less space to spread out.[67] At a flood conference in Calcutta in 1897, an army officer compared embankments to throwing one's gloves. By the late colonial period, engineers in Bihar had almost started giving up on embankments.[68] Captain GF Hall, the chief engineer of Bihar, told his colleagues in 1937 that "one point and one point only as a preliminary to government devising a flood policy, i.e., that bundhs obstruct the free flow of water and accentuate instead of relieve the intensity of floods".[69] Although opposed by the British engineers, the project was revived in post-independent India. The first Prime Minister of India Jawaharlal Nehru emphasised the importance of such a scheme. Referring to the strategy at the time of its initiation, Nehru stressed:

> [I]t is my opinion that the Kosi Project is very necessary and should be somehow constructed. We must make a start even though it may take a few years to complete because, as you know, in some parts of Bihar every year a strange difficulty arises, bringing disaster and ruin.[70]

In 1951, after the overthrow of the Rana oligarchy, a new government was installed in Nepal and more focused attention was directed to the Kosi project. The Central Water and Power Commission of the Government of India prepared a scheme for harnessing the Kosi River that received the sanction of the Government of India in 1953. Thereafter, the scheme was endorsed by Nepal's Government, following which the 1954 Kosi Agreement was negotiated and signed.[71] But soon after its conclusion, the 1954 Agreement was sternly criticised by the opposition political parties in Nepal. Critics asserted that the Project did not benefit Nepal in any manner whatsoever and that it granted extraterritorial rights to India for an indefinite period without providing Nepal with adequate compensation. It was also pointed that Nepal would receive only a minute proportion of the total irrigated land and India would benefit more from the power resources developed by Nepal.[72] Indian officials were criticised for interference in the internal affairs of Nepal and demands were raised for their removal from project work.[73]

The Nepali government was also criticised for its role. Defending the government, then Prime Minister of Nepal, M.P. Koirala said:[74]

> If one is determined to misunderstand a very plain situation, nobody ever can help him. India could have very well put the barrage a couple

of miles below the present agreed site if it had no consideration for Nepal. The sovereignty and territorial rights of Nepal had not been impaired by the Kosi agreement.

King Mahendra also defended the Kosi agreement. To address differences, the Kosi Project Coordination Committee was formed with officials from both India and Nepal as members. The Committee discussed matters such as acquisition of land, compensation, rehabilitation of displaced persons, maintenance of law and order, soil conservation, etc. and tried to iron out differences.[75] Later, the Indian power and irrigation minister, Dr K.L. Rao, visited Nepal during 1962–63. At that time India showed willingness to amend the Kosi agreement in light of the complaints made by Nepal [76] but it did not fully convince the Nepali government which told India to stop work on the western canal system pending further discussions on the revision of the agreement. The work on project restarted only in 1965 when the then Prime Minister of India, Lal Bahadur Shastri during his visit to Nepal, assured the government about the amendment to the Kosi agreement.[77]

The Kosi project as originally conceived was to be a 750-foot-high dam at Barahakshetra in Nepal with a storage capacity of 11 million-acre feet and 1200 MW power station. The height was subsequently lowered down to 85-feet high earthen dam at Belka hills, nine miles downstream from the proposed high dam site.[78] The smaller project was designed to generate only 68 MW of power and irrigate 1.52 million acre-feet. The scheme, which was finally implemented, consisted of only a barrage and embankments and a ring bund essentially as a flood protection measure.[79]

In 1966, the Kosi treaty was amended. In the amended treaty it was clarified by Nepal that

> that the Government of India will be reasonably compensated in case the Project properties are taken over by His Majesty's Government at the end of the lease period. The compensation will cover the cost borne to date and such other cost as may be incurred in future by the Government of India with the agreement of His Majesty's Government. In that case the depreciation in the value of the Project materials would, of course, be taken into account.[80]

In 1978 India agreed to finance yet another project to overcome Nepali resentment.[81] A sum of 1805 lakh (US $0.23 million) of Nepali Rupees was made available to Nepal for it to undertake the renovation and extension of the Chandra Canal, its distribution system, and the related works to provide irrigation for net command area of 34,690 hectares in Nepali territory.[82]

Despite projections, the projects on the River Kosi could not protect the people in the catchment areas from the floods. The project which had been designed to protect over 2,000 square kilometres of land water-logged over 3,000 square kilometres by 1990s.[83]

Floods are regular in the Kosi River. In recent times, in 2008 the catchment regions of Kosi in Bihar and Nepal witnessed severe floods in which hundreds of people died and a number of them were displaced. One of the reasons for the continuous flooding in the region is siltation. That year Kosi breached its embankment in Kushaha in Nepal and then shifted 108 kilometres eastwards. At that time, it was estimated that around 1082 million tonnes of silt was deposited between Chhatra and Birpur (two gauge stations of Kosi) in the post-embankment period.[84]

In successive years again floods have occurred in Kosi River due to rain and release of waters in Nepal. On July 11–12, 2019, Nepal's Simara weather station received more than half the 580.2-millimetre rainfall it normally gets in July during the monsoon.[85] As a result of the spell of 478.40 mm, all 56 sluice gates of the Kosi barrage were opened. It released about three lakh cusecs towards Bihar. This has caused floods in Kosi, Bagmati, Kamka Balan, Gandaki, Budhi Gandaki, and their tributaries. The floods had affected more than 2.5 million people.[86] To manage the floods, in 2019, the Union government approved the Indian Rupees 4,900 crore (US $6.498 billion) project for the interlinking of Kosi and Mechi Rivers of Bihar. Under it, the central government has approved construction of 76.20 kilometres of canals on the eastern bank of Kosi for irrigation purposes.[87] It is assumed that the project will not only prevent recurring floods in north Bihar, but also irrigate over 2.14 lakh hectares of cultivable land in Araria, Purnea, Kishanganj, and Katihar districts which are together called Seemanchal region. This is a "national project" because the entire command area of the rivers lies in both India and Nepal.[88]

In 2018, India has also provided an assistance of about 167.62 million Nepali Rupees (US $13.8 million) for river training and construction of embankments along the Lalbakeya, Bagmati, and Kamla Rivers in Nepal. It is being aimed at flood control and water resources management, which benefit several million people inhabiting in the watershed of these rivers in India and Nepal.[89] This amount is a part of Nepali Rupees 4.85 billion (US $4.12 million) instalment that India has committed to provide Nepal for further cooperation in the field of river training, flood control, and water resources management in the mentioned and rivers.[90]

Gandaki river water treaty

Efforts toward harnessing the large irrigation potential of the River Gandaki (or Gandak in India) had been made as early as 1871. Formally, it was initiated in 1947 with the construction of a canal in Tribeni.[91] In 1947, Dr Rajendra Prasad, the then Food and Agriculture Minister of India, wrote to the Government of Bihar to explore the possibilities of constructing a canal system from the Gandaki for irrigation. In 1951 a report was prepared in this connection and submitted to the Planning Commission of

Figure 4.1 River Gandaki, Sitalpur, Saran, Bihar
Source: Photo by author

India, which accepted the proposal and forwarded to the Government of Nepal who also endorsed it, and, in December 1959, an Agreement was concluded.[92] The Gandaki Project consists of construction of a barrage, canal head regulators, and other appurtenant works about 1,000 feet below the existing Tribeni canal head regulator. The Project was built by India at the cost of Indian Rupees 50.5 crores (US $6.16 million).[93]

Although not completely immune from shortcomings and criticisms, the provisions of the Gandaki Agreement, from Nepal's perspective, looked relatively favourable with those of the Kosi Agreement.[94] For instance, while Article 4 of the Kosi Agreement gave Nepal 50% share of the hydroelectric power that India generated as a result of the Project, it did not require India to produce any power.[95] In fact, India has not generated any such power, thus depriving Nepal of the promised benefit. Again Article 10 of the Kosi Agreement stipulated that provision shall be made for free and unrestricted navigation at and around the barrage "if technically feasible", the qualifying conditional words providing a justification for India's doing nothing about navigation as required by the Article.[96] As against 1.60 million hectares it was designed to irrigate in UP and Bihar states of India, it only planned to provide irrigation for 15,800 hectares in Nawalparasi district of Nepal. The canals under the project were aligned in such a manner that it led to water logging in large areas of Nepal.[97]

There were mixed reactions in Nepal over Gandaki agreement, but it was criticised by a sizeable population. The construction of barrage in Nepali territory was looked at as Indian encroachment on Nepal's sovereignty and territorial integrity.[98] The Indian government was accused of pressuring the Nepali government to conclude the agreement when the king was out of the country. It was argued by a number of political commentators in Nepal that before signing the agreement, the government should have sought advice of experts from a third country like Sweden or Switzerland to ascertain that it had done no mistake by concluding the Gandaki agreement.[99] Criticising the agreement, T. P. Acharya, former Prime Minister of Nepal, said:[100]

> It is necessary for Nepal to work in collaboration with India for executing the Gandak project. But it is evident that Nepal will have very little benefit from this project, at the cost of giving its land to the possession of another country. I request, therefore, that Nepal Government should think for its own benefits also. I also request the Indian government not to force the present weak Government to take any hasty decision on such a (n) important topic.

Eventually, in April 1964 the Gandaki treaty was amended. The exchange of letters relating to 30 April 1964 deleted clause 10 and clause 9 was modified to read as:

> His Majesty's Government will continue to have the right to withdraw for irrigation or any other purpose from the river or its tributaries in Nepal such supplies of water as may be required by them from time to time in the Valley. For the trans-Valley uses of Gandaki waters, separate agreements between His Majesty's Government and the Government of India will be entered into for the uses of waters in the months of February to April only.[101]

Clause 7 (v) of the treaty includes "Also, the head regulator of the Don branch canal shall be operated by His Majesty's Government keeping in view the irrigation requirements of areas irrigated by this branch canal in India and Nepal".[102]

India-Nepal water relations are beyond treaties signed on major rivers like Kosi, Gandaki, Karnali, or Mahakali to harness hydroelectric and for irrigation projects. According to an estimate, if one counts rivers, streams, and watercourses, about 6,000 such water bodies are shared between India and Nepal.[103] Some of the smaller rivers between India and Nepal are tributaries to major Indian rivers like the Ganga, etc. For example, the River Pandai which flows along the eastern border of the Valmiki Tiger Reserve once it enters India, and then within West Champaran, Gopalganj, Saran and Muzaffarpur districts of Bihar before joining the Masan and then River

Ganges. It ultimately becomes part of the Ganga-Brahmaputra-Meghna basin on which about 630 million people in Nepal, Bhutan, India, and Bangladesh depend.[104]

Nepal: drifting away from India

To safeguard the waters of Nepal, in 1977 the King of Nepal Briendra Bir Bikram Shah put forth the concept of regional cooperation in the development of significant Nepali water resources.[105]

In order to break the Indian stranglehold on its waters, Nepal for some time toyed with the idea of regional cooperation for utilising the waters of the Himalayan Rivers by bringing in Bangladesh and China.[106] Nepal wanted to do so because it had a different strategy to use its water resources: While India's interest in Nepali waters was for consumptive purposes, besides generation of power and flood control, Nepal was more interested in the navigational aspect for easy and cheaper access to the sea through Bangladesh's territory.[107]

Over water issues, as early as in 1978 then Nepali Prime Minister Kirtnidhi Bista in New Delhi, in a Joint Communiqué on 17 April 1978 said that Nepal would be happy if China could participate in the regional development of water resources and could spare some finances for such development.[108] One of the complaints Nepal has against India is that the projects carried out by India has led to the inundation of its land. An example of this is The Lothal-Raisawal-Khurd bund. Nepal noted that due to it there was inundation in and around the Marcheshwar area of Pupondehi in Nepal. The issue was taken to the High Level Technical Committee (HLTC) set up to look into the matter. But instead of looking into the matter the HLTC formed a team and committees for joint field verification.[109] The problems remain unresolved for many years. Prior to it, in 2000 rehabilitation works including replacement of wooden gates with steel ones in some of the structures without prior information led to the inundation of land in Kapilvastu area where there are many sagars (small water bodies)on the India-Nepal border.[110]

To give up its dependence on India, Nepal tried to develop water links with Bangladesh. The question of Nepal's participation in the India-Bangladesh talks on Ganga waters was discussed in November 1982 between Nepal and Bangladesh during the visit of then President of Bangladesh H.M. Ershad to Kathmandu. Ershad said he would devote his "best efforts" in talks with the Indian leaders on a proposal to link Bangladesh and Nepal through a 29 kilometres strip of Indian territory.[111] The water link was expected to give Nepal an outlet to the sea by linking Nepal's river system with the Atrai River in Bangladesh. But he did not make any serious effort in this direction.[112] In 1986, at Dhaka's insistence to involve Nepal to study the feasibility of storages on the Ganges for the augmentation of the Ganges flows,

New Delhi agreed to discuss the issue in a tripartite meeting. The meeting of the experts from the three countries took place in October 1986. But it could not reach the solution of the augmentation problem.[113]

In recent years, Nepal has pushed itself to explore options with China. One of the recent developments that made such tilt possible was the economic blockade of Nepal by India from September 2015 to February 2016 which created energy shortage in Nepal.[114] Afterwards, China has made strong inroads in Nepal.

Since 2016, China has made investments in Nepal's security forces, particularly in its police and paramilitary forces. It even opened up an academy to train Armed Police Force and deployed the Nepali paramilitary force to guard the Nepal-Tibet border from possible infiltration of "Free Tibet" activists into Nepal.[115] China and Nepal also started conducting joint military exercises. The first one called *Sagarmatha* (Ocean Mother) was conducted in 2017 in Kathmandu. A second joint exercise was held in Chengdu in 2018.[116] This sort of friendly relationship is also making the two countries to engage more in the water sector.

Earlier, in November 2017, the Nepali government, allegedly, under India's pressure had to cancel a US $2.5 billion deal with China's Gezhouba Group to build the Budhi Gandaki Hydroelectric Project.[117] This decision was overturned after K.P. Sharma Oli became Prime Minister in February 2018. In September 2018, K. P. Oli's government handed over this 1200 MW project to the same Chinese company.[118]

In in June 2018, during the K.P. Oli's visit to China, the two sides expressed

> willingness to speed up the development of the three North-South Economic Corridors in Nepal, namely Koshi Economic Corridor, Gandaki Economic Corridor and Karnali Economic Corridor in order to create jobs and improve local livelihood, and stimulate economic growth and development in those areas.[119]

Nepal had also invited the Chinese state-run Three Gorges Corporation to build a dam on River Seti in Western Nepal. The Chinese firm wanted 75% stake in this project.[120] In 2018 the negotiations between the China and Nepal failed; as a result, the China pulled out from this project. In June 2019, Chinese company Sinohydro Corporation Limited was employed to complete the remaining works related to tunnel works of Melamchi Water Supply Project, Nepal's largest water supply project. Earlier the project was with an Italian contractor *Cooperativa Muratori e Cementisti di Ravenna*. This project aims to supply 170 million litres of fresh water per day to capital Kathmandu.[121] Sinohydro has been involved in other Nepali projects as well. It has jointly developed 50MW Upper Marsyangdi Hydropower Project with Nepali company Sagarmatha

Power Company. Sinohydro is also a contractor for under-construction 456MW Upper Tamakoshi Hydropower Project, the largest domestically funded hydroelectricity project in Nepal and 140MW Tanahu Hydroelectricity project.[122]

China has also built a tunnel for water diversion which will transfer water from the Bheri River to the Babai River and provide irrigation for 51,000 hectares of farmland in central and western Nepal.[123] In 2019 during the visit of the Chinese President Xi Jinping's visit to Nepal, the two countries reiterated the MoU they signed in 2018 over further cooperation on hydropower sector.

Nepal's Woes of Having in Plenty

Although a water rich country, many in Nepal feels that their resources are not being used for their benefit. Some also feel that they are being exploited by different internal and foreign stakeholders who make enormous profits. To protect their water resources, Part 4 of the 2016 Nepal's constitution under Directive Principles, Policies and Responsibilities of the State has some provisions. Under Article 51 (g) the State shall pursue a policy:[124]

1   to make multi-purpose development of water resources, while according priority to domestic investment based on public participation,
2   to ensure reliable supply of energy in an affordable and easy manner, and make proper use of energy, for the fulfilment of the basic needs of citizens, by generating and developing renewable energy,
3   to develop sustainable and reliable irrigation by making control of water-induced disasters, and river management.

As these policies are under the Directive principles, policies and obligations of the State, they may not be viable enough to secure the local people's water, however, the State may take steps to implement them. Further, Article 59 (4) of the constitution states that

> The federation, State and the local level shall provide for the equitable distribution of the benefits from the use of natural resources or development of natural resources. Certain portions of such benefits, shall be distributed in areas, pursuant to law, in forms of royalty, services or goods to the project affected regions and local communities.[125]

Provision (5) of the same article says that

> if, in utilising natural resources, the local community desires to make investment therein, the Federation, State and Local level shall accord

priority to such investment in such portion as provided bylaw on the basis of the nature and size of such investment.[126]

A sizeable population of Nepal feels that despite being water rich they have not been benefitted a lot, from the HEP. To look at some of these issues, the case of Nepal was also studied by Benjamin Sovacool and Gotz Walter in their paper published, looking at the economic, environmental, and social impacts of hydropower.[127] They look at the issue of a "hydroelectric resource curse". In their paper, the authors argue that hydropower states have not dealt well with poverty, economic growth, public debt, and corruption. Nepal fits well to their study which they have indeed taken as one of the cases. After analyzing Sovacool and Walter's article, Eugene Simonov finds that hydroelectricity development in Nepal is uncoordinated which often leads to:[128]

- the construction of inferior projects, rather than more efficient designs on the same river;
- extreme fragmentation of river basins, decline in fish diversity, encroachment into protected areas and destruction of important wetlands;
- encroachment on areas traditionally used by indigenous peoples and marginalised groups;
- haphazard mass resettlement (the Budhi Gandaki dam alone will affect about 45,000 people);
- increased exposure to potential landslides, earthquakes and natural disasters combined with low resilience to climate change impacts;
- conflict with other sectors such as agriculture, fisheries, water supply, municipal wastewater, navigation, mining, etc.; and
- financial irregularities, increased corruption and investment schemes with high probability of failure.

There have been displacements due to HEP in Nepal but, so far, its magnitude is not as large as in some other countries, such as India, China, Sri Lanka, and Indonesia. However, it is expected to increase in coming years with the construction of many more HEP projects which are in the pipeline.[129] There are policies and laws to compensate the affectees of the HEP projects. However, in most cases, they are not being appropriately compensated.

The growing ill-effects of HEP projects have made many civil-society members and activists raise questions on them. There is a growing, what Sanju Koirala calls, hydro activism, in Nepal. These hydro activists are particularly against engaging a large number of Indian companies in Nepal's hydropower sector. They argue that India has always cheated the Nepalis on terms and conditions of HEP projects.[130] There were protests against the works on the Karnali and Kosi high dam due to which their construction work has been delayed.[131]

Conclusion

This chapter discussed the India-Nepal transboundary water sharing issues, in light of political relationship between the two countries. It has been maintained that both water sharing arrangements with India and political relationships have influenced each other. Nepal always accuse India of exploiting its water resources for its own benefit. This is the reason why most of their bilateral water sharing treaties have been either re-negotiated or not been activated or have taken a lot of time to come into effect. This has given rise to anti-India constituency in Nepal which has been politically exploited by some political parties. While discussing the issue with people from Nepal, the author found that a number of them were convinced that they do not get the benefits of their waters. They were also supportive to the idea of having China in the region so that India's monopoly could be checked.

To mend its relations with Nepal, in recent times India has entered into several agreements to improve infrastructure in Nepal. On 21 August 2019, S. Jaishankar visited Nepal to attend the fifth meeting of the Nepal-India Joint Commission. During the meeting, both sides reviewed their bilateral relations with specific focus on the areas of connectivity and economic partnership; trade and transit; power and water resources sectors; culture; and education. The India delegation presented a cheque worth Nepali Rupees 2.45 billion (US $20.26 million) to Nepal as reimbursement for housing reconstruction in Nuwakot and Gorkha districts. Indian side also gave a cheque worth Indian Rupees 80.71 crore (US $0.107 million) to Nepal, as part of INR 500 crore (US $66 million) to strengthen road infrastructure in the Nepal's Terai Region.[132]

As discussed in the chapter, every year transboundary rivers from Nepal cause floods in Bihar. To deal with the situation the two governments have set up mechanisms and built infrastructures. However, all such mechanisms and infrastructures have failed to manage such regular floods. In 2008 Bihar witnessed the worst floods when the Kosi River changed the track. Aftermath measures were taken to manage such floods but in 2014, 2015, 2017, and 2019 the same area witnessed floods. After 2019, the Indian government accepted the proposal to link Mechi River with Kosi. It is being estimated that this would manage flood and help in irrigation. However, its feasibility to fulfil both objectives has to be observed.

Notes

1 Ministry of External Affairs, Government of India, Indian Embassy in Nepal. "About Trade and Commerce". www.indembkathmandu.gov.in/page/about-trade-and-commerce/. Accessed on 25 August 2019.
2 Baral, Toya Nath Border Disputes and Its Impact on Bilateral Relation: A Case of Nepal- India International Border Management https://www.nep jol.info/index.php/JAPFCSC/article/download/26710/22113/. Accessed on 18 November 2019.

3 Constitution of Nepal. (2015). "Constituent Assembly Secretariat". www.icnl. org/research/library/files/Nepal/Nepalconst.pdf, p. 148.
4 Subedi, P. Subedi. (2005). *Dynamics of Foreign Policy and Law: A Study of Indo-Nepal Relations.* New Delhi: Oxford University Press, p. 120.
5 England, Mathew I. and Daniel Haines. (2018). "Topography and the Hydraulic Mission: Water Management, River Control and State Power in Nepal". *International Journal of Water Resources Development.* https://doi.org/10.1080/079 00627.2018.1515066.
6 Government of Nepal, Ministry of Foreign Affairs. "Nepal-India Relations". https://mofa.gov.np/nepal-india-relations/. Accessed on 14 July 2019.
7 Ibid.
8 Ibid.
9 Cited in England, Mathew I. and Daniel Haines. (2018). "Topography and the Hydraulic Mission: Water Management, River Control and State Power in Nepal". *International Journal of Water Resources Development.* https://doi.org /10.1080/07900627.2018.1515066.
10 Ibid.
11 Parikh, Kirit. (2017, 13 April). "Powering India-Nepal Ties". *The Hindu.* www.thehindu.com/opinion/lead/powering-india-nepal-ties/article17960821. ece?homepage=true. Accessed on 13 April 2017.
12 Ibid.
13 England, Mathew I. and Daniel Haines. (2018). "Topography and the Hydraulic Mission: Water Management, River Control and State Power in Nepal". *International Journal of Water Resources Development.* https://doi.org/10.1080/079 00627.2018.1515066.
14 Salman, M.A. and K. Uprety. (2002). *Conflicts and Cooperation on South Asia's International Rivers: A Legal Perspective.* Washington, DC: The World Bank.
15 "1920 Sarada Barrage Project Assessment Between British India and Nepal". *International Rivers.* www.internationalrivers.org/sites/default/files/attached-files/treaties_between_nepal-india.pdf. Or http://nepaldevelopment.pbworks. com/w/page/34196794/Nepal-India%20Relations. Accessed on 18 June 2018.
16 Cited in Upreti, B.C. (1993). *Politics of Himalayan River Waters: An Analysis of the River Water Issues of Nepal, India and Bangladesh.* Jaipur: Nirala Publications, p. 94.
17 Siddiqi, Toufiq A. and Shirin-Tahir-Kheli. (2004). Water Conflicts in South Asia: Managing Water Resource Disputes Within and Between Countries of the Region, Project Implemented by Global Environment and Energy in the 21st Century (GEE-21) and the School of Advanced International Studies, John Hopkins University (SAIS), Sponsored by the Carnegie Corporation of New York, p. 163.
18 Salman, M.A. and K. Uprety. (2002). *Conflicts and Cooperation on South Asia's International Rivers: A Legal Perspective.* Washington, DC: The World Bank.
19 Memorandum of Understanding (MoU) on Tanakpur Barrage Project, 1991 *See* the link www.internationalrivers.org/files/. . ./treaties_between_nepal-india.pdf.
20 Iyer, Ramaswamy. (2002). "Three Waters Treaty". In Sahadevan, P. (ed) *Conflicts and Peacemaking in South Asia.* New Delhi: Lancers Publication, pp. 365–395.
21 Ibid.
22 Gyawali, Dipak. (2007). *Water in Nepal.* Kathmandu: Himal Books, p. 54.
23 Bhasin, A.S. (2005). *Nepal-India, Nepal-China Relations Documents 1947-June 2005 Vol-I.* New Delhi: Geeta Press, p. xlviii.
24 Ministry of Jal Shakti, Department of Water Resources, River Development & Ganga Rejuvenation, Government of India Mahakali Treaty 1996, http://mowr. gov.in/sites/default/files/MAHAKALI_TREATY_19961.pdf. Accessed on 18 July 2019.

25 Ibid.
26 Ibid.
27 Ministry of Jal Shakti, Water Resources River Development & Ganga Rejuve-
    nation, Government of India. "India-Nepal Cooperation". http://mowr.gov.in/
    international-cooperation/bilateral-cooperation-with-neighbouring-countries/
    india-nepal-cooperation.
28 Ibid.
29 Ibid.
30 Ibid.
31 Ibid.
32 Ibid.
33 Ministry of Jal Shakti, Water Resources, River Development & Ganga Reju-
    venation "Pancheshwar Multipurpose Project". http://mowr.gov.in/schemes-pro
    jects-programmes/projects/pancheshwar-multipurpose-project. Accessed on 14
    February 2019.
34 "India, Nepal to Fast-track Pancheswar Development Authority". *The Times of
    India*, 2002, 16 February. https://timesofindia.indiatimes.com/india/India-Nepal-
    to-fast-track-Pancheshwar-Development-Authority/articleshow/11905123.cms.
    Accessed on 12 July 2018.
35 "Revised DPR for Pancheshwar Dam Sent to Project Development Authority".
    Outlook, 2018, 04 September. https://www.outlookindia.com/newsscroll/revised-
    dpr-for-pancheshwar-dam-sent-to-project-development-authority/1376033.
    Accessed on 17 September 2018.
36 Jha, Prashant. (2012, 5 September). "Nepal Maoist Serve Ultimatum". *The Hindu*.
    https://www.thehindu.com/news/international/nepal-maoist-faction-serves-
    ultimatum/article3863011.ece. Accessed on 18 October 2107.
37 Asher, Manshi. (2018, 19 February). "What Lies Behind the Resistance to
    India's Highest Dam?" *The Wire*. https://thewire.in/environment/uttarakhand-
    lies-behind-resistance-building-countrys-highest-dam. Accessed on 19 February
    2019.
38 Ibid.
39 Ibid.
40 Chakrabarty, Arpita and Prem Punetha. (2017, 23 July). *The Times of India*.
    https://timesofindia.indiatimes.com/city/dehradun/villagers-protest-con
    struction-of-pancheshwar-dam/articleshow/59716912.cms. Accessed on 24 July
    2017.
41 Asher, Manshi. (2018, 19 February). "What Lies Behind the Resistance to
    India's Highest Dam?" *The Wire*. https://thewire.in/environment/uttarakhand-
    lies-behind-resistance-building-countrys-highest-dam. Accessed on 19 February
    2019.
42 Ibid.
43 Iyer, Ramaswamy R. (2008, 17 September). "Water in India-Nepal Relations".
    The Hindu. https://www.thehindu.com/todays-paper/tp-opinion/Water-in-India-
    Nepal-relations/article15304717.ece#!. Accessed on 27 June 2017.
44 "Cabinet Approves ₹1236 Crore Investment for Arun-3 Hydro Project". *Live
    Mint*, 2019, 1 March. www.livemint.com/politics/news/cabinet-approves-rs-
    1236-crore-investment-for-arun-3-hydro-project-1551381634214.html.
    Accessed on 16 May 2018.
45 Ministry of Foreign Affairs, Government of Nepal Annual Report. (2017–18).
    https://mofa.gov.np/wp-content/uploads/2019/02/Final-_-Mofa-Book.pdf.
    Accessed on 1 September 2019, p. 47.
46 Ministry of External Affairs, Government of India India-Nepal Joint Statement
    During the State Visit of Prime Minister of India to Nepal. (2018, 11–12 May).
    documents.htm?dtl/29894/IndiaNepal_Joint_Statement_during_the_State_

Visit_of_Prime_Minister_of_India_to_Nepal_May_1112_2018. Accessed on 12 March 2019.

47 Ibid.

48 "Cabinet Approves ₹1236 Crore Investment for Arun-3 Hydro Project". *LiveMint*, 2019, 1 March. www.livemint.com/politics/news/cabinet-approves-rs-1236-crore-investment-for-arun-3-hydro-project-1551381634214.html. Accessed on 12 June 2017.

49 Ibid.

50 Koirala, Sanju. (2015). "Hydropower Induced Displacement in Nepal". Unpublished PhD Thesis, Submitted at University of Otago Dunedin, New Zealand. https://otago.ourarchive.ac.nz/bitstream/handle/10523/6141/KoiralaSanju2016PhD.pdf?sequence=3&isAllowed=y. Accessed on 25 July 2019, p. 89.

51 Upreti, B.C. (1993). *Politics of Himalyan River Waters: An Analysis of the River Water Issues of Nepal, India and Bangladesh*. Jaipur: Nirala Publications.

52 Ibid.

53 Prasin, Sangam (2020, 9 February) Bangladesh issues letter of intent to purchase 500 MW from Upper Karnali hydro project.*The Kathmandu Post*. https://kathmandupost.com/money/2020/02/09/bangladesh-issues-letter-of-intent-to-purchase-500-mw-from-upper-karnali-hydro-project. Accessed on 20 February 2020.

54 "GMR Signs Agreement with Nepal for 900MW Hydro Project". *Live Mint*, 2014, 23 September. www.livemint.com/Companies/Pu8KypRY2WXHRR8ZzoUTYP/GMR-signs-agreement-with-Nepal-for-900MW-hydro-project.html. Accessed on 15 July 2018.

55 "Nepal's Supreme Court vacates stay order against GMR project " *The Economic Times*, 2013, 12 August. https://economictimes.indiatimes.com/industry/indl-goods/svs/construction/nepals-supreme-court-vacates-stay-order-against-gmr-project/articleshow/21782886.cms. Accessed on 19 July 2018.

56 "Petrol bomb hurled at GMR Group's office in Kathmandu " *The Economic Times*. 2016, 5 January. https://economictimes.indiatimes.com/news/politics-and-nation/petrol-bomb-hurled-at-gmr-groups-office-in-kathmandu/articleshow/50458864.cms. Accessed on 18 July 2017.

57 Prasin, Sangam (2020, 9 February) Bangladesh issues letter of intent to purchase 500 MW from Upper Karnali hydro project.*The Kathmandu Post*. https://kathmandupost.com/money/2020/02/09/bangladesh-issues-letter-of-intent-to-purchase-500-mw-from-upper-karnali-hydro-project. Accessed on 20 February 2020.

58 Upreti, B.C. (1993). *Politics of Himalyan River Waters: An Analysis of the River Water Issues of Nepal, India and Bangladesh*. Jaipur: Nirala Publications.

59 Ibid.

60 Chauhan, Chetan and Utpa Prashar. (2016, 10 February). "India's Thirst for Water May Address Power Woes in Nepal". *The Hindustan Times*. www.hindustantimes.com/india/india-s-thirst-for-water-may-address-power-woes-in-nepal/story-pVeRGwKHxxtD9lfbs8eRYI.html. Accessed on 9 February 2016.

61 Ibid.

62 Ibid.

63 Ibid.

64 Ministry of External Affairs, Government of India. "India-Nepal Statement on New Connectivity Through Inland Waterways". https://mea.gov.in/bilateral-documents.htm?dtl/29796/IndiaNepal_Statement_on_New_Connectivity_through_Inland_Waterways. Accessed on 24 June 2018.

65 Ibid.

66 Gill, Peter and Bhola Paswan. (2018, 15 August). "A Dammed History of the Koshi". *thirdpole.net*. https://www.thethirdpole.net/2018/08/15/a-dammed-history-of-the-koshi-in-bihar/. Accessed on 17 August 2018.
67 Ibid.
68 Ibid.
69 Cited in Ibid.
70 Cited in Salman, M.A. and K. Uprety. (2002). *Conflicts and Cooperation on South Asia's International Rivers: A Legal Perspective*. Washington, DC: The World Bank, p. 66.
71 Ibid.
72 Ibid.
73 Upreti, B.C. (1993). *Politics of Himalyan River Waters: An Analysis of the River Water Issues of Nepal, India and Bangladesh*. Jaipur: Nirala Publications.
74 Ibid., p. 99.
75 Ibid.
76 Ibid.
77 Ibid.
78 Bhasin, A.S. (2005). *Nepal-India, Nepal – China Relations Documents 1947-June 2005 Volume II*. New Delhi: Geeta Press, p. Xliii.
79 Ibid.
80 Ministry of Economic Affairs, Government of Nepal. (1966). www.internationalrivers.org/sites/default/files/attached-files/treaties_between_nepal-india.pdf.
81 Bhasin, A.S. (2005). *Nepal-India, Nepal – China Relations Documents 1947-June 2005 Volume II*. New Delhi: Geeta Press, p. xlvi.
82 Ibid.
83 Gill, Peter and Bhola Paswan. (2018, 15 August). "A Dammed History of the Koshi". *thirdpole.net*. https://www.thethirdpole.net/2018/08/15/a-dammed-history-of-the-koshi-in-bihar/. Accessed on 17 August 2018.
84 Kaur, Banjot. (2018, 3 April). "Why Does Kosi River Cause Devastating Floods So Often? Answer Lies in Massive Siltation: Study". *Down to Earth*. www.downtoearth.org.in/news/water/why-does-kosi-river-cause-devastating-floods-so-often-answer-lies-in-massive-siltation-study-60014. Accessed on 12 May 2018.
85 Khan, Mohd Imran. (2019, 18 July). "Bihar Is Flooding, But Where Did It Start? Hint: Look North". *Down to Earth*. www.downtoearth.org.in/news/natural-disasters/bihar-is-flooding-but-where-did-it-start-hint-look-north-65683. Accessed on 25 July 2019.
86 Ibid.
87 "Bihar Gets Centre's Nod for Kosi-Mechi River Interlinking". *Business Standard*, 2019, 3 August. www.business-standard.com/article/pti-stories/bihar-gets-centre-s-nod-for-kosi-mechi-river-interlinking-119080300939_1.html. Accessed on 4 August 2019.
88 Ibid.
89 Ministry of External Affairs, Government of India Indian Embassy in Nepal. "About Trade and Commerce". www.indembkathmandu.gov.in/page/about-trade-and-commerce/. Accessed on 25 August 2019.
90 Ibid.
91 Salman, M.A. and K. Uprety. (2002). *Conflicts and Cooperation on South Asia's International Rivers: A Legal Perspective*. Washington, DC: The World Bank.
92 Salman, M.A. and K. Uprety. (2002). *Conflicts and Cooperation on South Asia's International Rivers: A Legal Perspective*. Washington, DC: The World Bank, p. 83.
93 Ibid., p. 84.

94 Ibid.

95 Ibid.

96 Ibid.

97 Bhasin, A.S. (2005). *Nepal-India, Nepal – China Relations Documents 1947-June 2005 Volume II*. New Delhi: Geeta Press.

98 Upreti, B.C. (1993). *Politics of Himalyan River Waters: An Analysis of the River Water Issues of Nepal, India and Bangladesh*. Jaipur: Nirala Publications.

99 Ibid.

100 Ibid., p. 105.

101 Ministry of External Affairs, Government of India. "Exchange of Letters Relating to Gandak Project". https://mea.gov.in/bilateral-documents.htm?dtl/6374/Exchange+of+Letters+relating+to+Gandak+Project. Accessed on 12 July 2018.

102 Ibid.

103 Siddiqui, Shawhaiq. (2017, 6 May). "India-Nepal Border Plagued by Water Trobules". *The Wire*. https://thewire.in/132733/india-nepal-border-plagued-water-troubles/ Accessed on 11 May 2017.

104 Ibid.

105 Dixit, A. (1997). "Regionalism Washed Away". *Himal Magzine*. http://old.himalmag.com/component/content/article/2634-Regionalism-Washed-Away.html. Accessed on 21 December 2015.

106 Bhasin, A.S. (2005). *Nepal-India, Nepal – China Relations Documents 1947-June 2005 Volume II*. New Delhi: Geeta Press, p. xlvii.

107 Ibid.

108 Ibid., p. 696.

109 Dhungal, Dwarika Nath, Santa Bhadur Pun and Basistha Raj Adhikari. (2009). "Inundation at Southern Border". In Dhungal, Dwarika Nath, Santa Bhadur Pun and Basistha Raj Adhikari (eds) *The India Nepal Water Relationship*. Amsterdam: Springer, pp. 269–294.

110 Ibid.

111 Bhasin, A.S. (2005). *Nepal-India, Nepal – China Relations Documents 1947-June 2005 Volume II*. New Delhi: Geeta Press.

112 Ibid.

113 Ibid.

114 Swain, Ashok. (2018, 7 April). "It Is Water Not China That Has Ruined Nepal's Relations with India". *Outlook*. www.outlookindia.com/website/story/it-is-water-not-china-that-has-ruined-nepals-relations-with-india/310684.

115 Baral, Biswas. (2017, 1 February). "After the 'Blockade': China's Push into Nepal". *The Diplomat*. https://thediplomat.com/2017/02/after-the-blockade-chinas-push-into-nepal/. Accessed on 25 August 2019.

116 Rajgopalan, Rajeshwari Pillai. (2018, 20 August). "Should Rising China-Nepal Military Ties Worry India?" *The Diplomat*. https://thediplomat.com/2018/08/should-rising-china-nepal-military-ties-worry-india/. Accessed on 22 August 2019.

117 Swain, Ashok. (2018, 7 April). "It Is Water Not China That Has Ruined Nepal's Relations with India". *Outlook*. www.outlookindia.com/website/story/it-is-water-not-china-that-has-ruined-nepals-relations-with-india/310684. Accessed on 12 April 2018.

118 News 18 "Nepal to Award 1,200 MW Hydro Project to Chinese Company". 2018, 4 September. www.news18.com/news/world/nepal-to-award-1200-mw-hydro-project-to-chinese-company-1887953.html. Accessed on 6 September 2018.

119 Ministry of Foreign Affairs, People's Republic of China. (2018, 22 June). "Joint Statement Between the People's Republic of China and Nepal". www.fmprc.

gov.cn/mfa_eng/wjdt_665385/2649_665393/t1570977.shtml. Accessed on 18 July 2018.
120 Chellaney, Brahma. (2017, 25 January). "Beijing Is Steadily Increasing Its Clout in Nepal at India's Expense". *Hindustan Times*. https://www.hindusta ntimes.com/analysis/beijing-is-steadily-increasing-its-clout-in-nepal-at-india-s-expense/story-aK7PYsIEUi9QBFff084OAO.html. Accessed on 23 July 2019.
121 *Xinhua.* (2019, 30 September). "Chinese Company Signs Contract to Develop Nepal's Largest Water Supply Project". http://www.xinhuanet.com/english/2019-09/30/c_138437365.htm. Accessed on 5 October 2019.
122 Ibid.
123 Qingyun, Wang. (2019, 19 April). "Nepal Tunnel Completed Ahead of Schedule Exemplifies BRI Cooperation, Envoy Says". *China Daily*. http://www.chi nadaily.com.cn/a/201904/19/WS5cb99bcba3104842260b73b7.html. Accessed on 12 June 2019.
124 Nepal Law Commission, Constitution of Nepal. (2015). http://www.lawcom mission.gov.np/en/. Accessed on 25 February 2019.
125 Ibid.
126 Ibid.
127 Sovacool, Benjamin and Gotz Walter. (2019). "Internationalizing the Political Economy of Hydroelectricity: Security, Development and Sustainability in Hydropower States". *Review of International Political Economy*, Vol. 26, No. 1, pp. 49–79.
128 Simonov, Eugene. (2019, 16 January). "Nepal's Hydropower Boom Needs Strategic Assessment and Public Oversight". *The Thirdpole.net*. www.thethird pole.net/en/2019/01/16/nepals-hydropower-boom-needs-strategic-assessment-and-public-oversight/. Accessed on 25 June 2019.
129 Koirala, Sanju. (2015). "Hydropower Induced Displacement in Nepal". Unpublished PhD Thesis, Submitted at University of Otago Dunedin, New Zealand. https://otago.ourarchive.ac.nz/bitstream/handle/10523/6141/Koirala Sanju2016PhD.pdf?sequence=3&isAllowed=y. Accessed on 25 July 2019.
130 Ibid., p. 89.
131 Swain, Ashok. (2018, 24 April). "Going Beyond the Mega Project Mindset Is a Must for India-Nepal Relations". *Down to Earth*. www.downtoearth.org.in/news/water/going-beyond-the-mega-project-mindset-is-a-must-for-india-nepal-relations-60285. Accessed on 12 July 2018.
132 "India Hands Over NPR 2.45 Billion to Nepal for Housing Reconstruction". *Business Standard*, 2019, 21 August. https://www.business-standard.com/article/news-ani/india-hands-over-npr-2-45-billion-to-nepal-for-housing-recon struction-119082101107_1.html. Accessed on 27 August 2019.

5 Concerns over India-assisted Hydroelectric Power Projects in Bhutan

Bhutan is one of the water-rich countries in the world. It has four major river system – Amo Chu, Wang Chu, Punatsang Chu, and Drangme Chu[1]– besides some small rivers like Jaldakha, Aiechhu, Nyera Amari, Jomori, and Merak-Sakteng. Bhutan has a total of 677 glaciers and about 2,794 glacial lakes.[2] Drangme Chhu (Manas) rises in the Himalayas and joins River Brahmaputra at Jogighopa in the Indian state of Assam. Second, Puntsang Chhu, after flowing southwards, enters Assam where it joins River Brahmputra. Third, Wang Chhu flows downstream to join Brahamputra. Amo Chhu, the smallest river system, rises in Tibet, flows through Western Bhutan, and then into India where it also joins River Brahmaputra. Due to a contiguous physical boundary, Bhutan shares many big and small rivers with India. It is estimated that about 56 small rivers flow down from Bhutan to the Indian state of Assam meet the Brahmaputra River.[3]

Bhutan receives around 70,576 million cubic metres of water every year.[4] This makes the per capita availability of water in the country more than 100,000 cubic metre.[5] However, this high per capita availability of water at the national level is in stark contrast to the growing local scarcity[6] in parts of Bhutan because of the phenomenon of climate change. In its study, carried out in 2016, the National Environment Commission, Royal Government of Bhutan found that some of the districts of the country like Thimphu, Haa, and Zhemgang may experience water shortage by the year 2030.[7] Furthermore, the study confirmed that due to the impact of the climate change, a few of the important water sources in the country have already started drying up.[8] It is being observed that due to the phenomenon of climate change, by 2015 Bhutan had lost 20% of its glaciers due to which river flows have been declined. Such decline in water flow is likely to increase if the impact of climate change continues. Melting glaciers also pose major flood risks, which could be catastrophic for dams and other hydropower projects.[9] Other than the climate change, people experience water scarcity because of inadequate water supply infrastructure in the country. Bhutan's topography makes difficult to set up infrastructure to provide regular water supply to the people scattered in the different parts of Bhutan.[10] There is also an issue of quality of water people consumes. It is estimated that despite availability only about 78% of the Bhutan's population has access to safe drinking water.[11]

Bhutan has a total capacity to generate about 30,000 megawatts (MWs) of hydroelectricity every year. Of which, only about 1,616 megawatts are being currently generated.[12] The 12th five-year plan (2018–2023) aims to enhance Hydroelectricity generation to about 2,444 MW by end of the plan period. The export of the hydroelectricity energy generates around 40% of the Bhutan's revenue. This sector's contribution to the country's Gross Domestic Product (GDP) is between 25 and 30%.[13] Some of the major commissioned, under-construction and proposed hydropower projects in Bhutan are:[14]

Hydroelectricity is centre to Bhutan's overall development and other economic sectors of the country are highly dependent on it. This has been unequivocally maintained in the official document titled *Bhutan 2020: A vision for peace, prosperity and happiness II*. This document was prepared by the Planning Commission of Bhutan and covers the period from 2016 to 2020. On the significance of Hydroelectricity *Bhutan 2020: A vision for peace, prosperity and happiness II* says:[15]

> Hydropower development . . . will generate the resources we require to maintain investments in the social services and the development of the physical infrastructure we require to continue to raise standards of living and the quality of life as well as to expand the level and pace of economic activity. Although we will use the cheap power produced to develop new resource-based processing industries in areas located close to their main export markets, we will have selected the theme of "sophistication and civilization" as the guiding principle for our industrial transformation. This theme will, two decades hence, find expression in the existence of clean industries based on a development-oriented interpretation of our resource endowments and comparative advantages and the existence of "high technology" enterprises, engaged in the production of high value/low volume products that place the nation in the vanguard of technological advance and innovation. Bhutanese high value products, some of which will be based upon the sustainable exploitation of the nation's rich biodiversity, will have a world market and will be recognized as the products of one of the least polluted and least contaminated countries to be found anywhere on earth.

Hydroelectricity's role in the country's economy has also been discussed in the Bhutan's *Twelfth Five Year Plan* (2018–2023). The plan report says that the hydroelectricity sector in Bhutan has been the main growth driver of the country through direct export earnings and its spillover effects on construction sector and power intensive industries.[16] The hydroelectricity sector alone has contributed an average of 16% in the country's GDP between 2007 and 2016. This will reach as high as 30% if Hydroelectric

Table 5.1 Hydroelectric Power Projects in Bhutan

Commissioned projects			Under construction			Proposed		
Name of project	Year of commission	Production capacity	Name of the project	Expected year of completion	Projected capacity	Name of the project	Proposed year	Expected capacity
Chuka	First unit 1986 and all three 1988	18,00 million units annually	Punatsangchhu-I	Last quarter of 2018	1200 megawatts	Bunakha	2013	180 megawatts
Kurichhu	2002–02	400 million units annualy	Punatsangchhu-II	Last Quarter of 2018	990–1020 megawatts	Amochhu	2013	520 megawatts
Basochhu	Upper stage in 2002 and lower stage in 2005	185 million units	Mangdechhu	Was to be completed by 2016 but completed in 2019	720 megawatts	Chamkharchhu	2015	2250 megawatts
Dagachhu	2015 restarted in December 2016	500,000 unit hours of power annualy	Wangchhu	Work started in 2014	570 megawatts	Sankosh	Expected to be completed by 2020	Revised to 2560 megawatts
Tala	2006	4865 gigawatt hours	Kholongchu	Foundation stone was laid in 2014	600 megawatts	Kuri Gongri	Report preparation began in 2014	Revised to 2640 megawatts
			Nikachhu	Environmental clearance in 2014	118 megawatts			

Source: Prepared by the author based on the following sources: Druk Green Power Corporation Limited, Government of Bhutan, www.drukgreen.bt/index.php/about-us & Status of Hydropower Dams in Bhutan International Rivers, 14 April 2015, www.internationalrivers.org/resources/8703, Department of Renewable Energy, Ministry of Economic Affairs, Government of Bhutan, www.moea.gov.bt/?page_id=1140

Power (HEP) projects construction is included.[17] The *Twelfth Five Year Plan* aims to enhance the total generation and also initiate efforts to develop HEP related services in the region.[18]

Most of the commissioned, ongoing, and proposed HEP projects have been funded by India because Bhutan lacks the required technology and finance. India's investment has been welcomed in past, however, in recent times a section of Bhutan's population is not very enthusiastic about it. For someone like political leader Pema Gyamtsho, HEP in Bhutan have turned into "a necessary evil" for a "resource-strapped country like Bhutan" to attain self-reliance.[19] It is being also argued that despite so much noise about them, Bhutan's HEP projects have not contributed substantially to improve the country's economy, as they have pushed the country into debt. For some critics the HEP's have not provided jobs to large number of Bhutanese or brought self-sufficiency in electricity generation.[20] This chapter discusses the India-Bhutan HEP sector and examines the concerns of Bhutanese against India's engagements in this important sector of their economy.

India-Bhutan hydroelectricity cooperation

In 1961, India and Bhutan signed one of the first HEP cooperation pacts. It was to harness hydroelectricity from the transboundary river, Jaldhaka.[21] Completed in 1966, the total capacity of the plant was around 18,000 kilowatts. According to the 1961 pact, Bhutan would receive the free supply of 250 kilowatts electricity from it.[22] In 1974, India and Bhutan signed another agreement which was the Chukha HEP project. This was the first mega power project fully funded by the Indian government with a 60% grant and a 40% loan. The interest rate on the loan was 5% payable over a period of 15 years. The capacity of this hydropower project was 336 megawatts.[23] The Chuka HEP project was fully commissioned in 1988.

Years after, in 2006, India and Bhutan signed a major agreement in the HEP sector. Under the 2006 Agreement, India and Bhutan agreed to "facilitate, encourage and promote the development and construction of hydropower projects and associated transmission systems as well as trade in electricity between the two countries, both through public and private sector participation".[24] An additional protocol to the 2006 agreement was signed in March 2009. Under the additional protocol to 2006, India has agreed to assist Bhutan in developing a minimum of 10,000 MWs of hydropower and import surplus electricity from the project to India by 2020.[25] In the protocol the two countries also agreed that "the modalities of developing this hydropower would be through existing models of direct Government of India assistance and through Indian and Bhutanese Public Sector Undertakings, in mutual consultation".[26] Earlier the projected capacity was of 5,000 MWs. It was first discussed to raise the capacity in 2008 after the visit by then Indian Prime Minister, Dr Manmohan Singh to Bhutan. During the visit India made a commitment to double the hydroelectricity capacity in Bhutan to 10,000 MWs by 2020. This was, mainly, to make Bhutan self-sufficient and reduce gap in trade deficits the country has with India.[27]

The format of the 2006 hydropower agreement is somewhat similar to the 1,020 MWs Tala HEP agreement between the two countries. The Tala hydroelectric project is the largest functional hydropower project in Bhutan. Work on this project began in 1998 and was finally commissioned in 2008. India's assistance for Tala came in the form of a 60% grant and 40% loan, with an interest rate of 9% per annum.[28] Indian companies Bharat Heavy Electrical Limited of India, Hindustan Construction Company, Larsen and Tubro, and Jaiprakash Industries participated in the Tala contract.[29] At present, India-assisted ongoing projects in Bhutan are:

> **Punatsangchu-I:** This is a 1,200-MW run-of-the-river project located on the left bank of the Punatsangchu River in Wangdue Phodrang Dzongkhag district in the Western parts of Bhutan. It is estimated that, once completed, the project is likely to generate around 5,700 million units of electricity every year. Its construction was started in 2008 and is expected to be completed by June 2019. The revised estimate of this project is around Indian Rupee 9,375.58 crore (US $12.4 billion). The loan-grant ratio for this project is 60:40.[30]
>
> **Punatsangchu-II:** This is also a run-of-the river project in Wangdue Phodrang Dzongkhag district. It is expected to produce around 4,357 million units of electricity every year. Its construction started in 2010 and is likely to be completed in the financial year 2018–19. The revised estimate for this project is Indian Rupees 7,290.62 crore (US $9.6 billion). The loan-grant ratio is 70:30.[31]
>
> Mangdechhu is another important 720-megawatt run-of-the-river hydroelectricity project. The estimated electricity generation from this project is around 2,925.25 million units every year. It was earlier expected to be completed by December 2018. The revised estimate of this project is Indian Rupees 4,020.63 crore (US $532 million) which was revised further to Indian Rupees 4672.38 (US $618 million) crore and finally settled at Indian Rupees 4500 crores (US $560 million). The loan-grant ratio is 70:30.[32] It was completed and inaugurated in 2019.

Besides these run-of-the-river projects, India and Bhutan are also cooperating in reservoir projects[33] such as Sankosh and Kuri-Gongri. The former has the capacity to generate about 4,060 MWs of hydroelectricity. Works began on Sankosh in 2011 and it is expected to be completed in 2020. Kuri-Gongri has a capacity to produce around 1,800 megawatts of hydroelectricity. Its construction started in 2012 and is expected to be completed in 2020.[34]

There are also projects under construction through the joint-venture model. Such projects are Kholongchhu, Chamkharchhu-I, Wangchhu, and Bunakha Reservoir.[35] The foundation stone of the run-of-the-river project,

Kholongchhu, in Trashiyangtse district of Bhutan, was laid during the visit of Indian Prime Minister Narendra Modi in 2014. The joint venture partners for this hydro projects are Druk Green Power Corporation of Bhutan and Satluj Jal Vidyut Nigam Ltd from India.[36] Its estimated cost is more than Rs 3,868 crores (US $512.29 million) which would be shared in the ratio of 50:50 by both the JV partners.

In February 2018, India's Foreign Secretary, Vijay Gokhale, while reporting on the status of India-assisted ongoing hydropower projects in Bhutan to the Parliamentary Committee on External Affairs (2018–19), stated:

> Bhutan has traditionally been the largest recipient of our [India] financial assistance for obvious strategic reasons that the Committee is familiar with. By June 2018, we will finish disbursing our commitment of INR 5,000 crore (US $662 million) for Bhutan's Eleventh Five Year Plan from 2013 to 2018. Indeed, we have already disbursed 80% of the funds exactly according to our implementation plan. About INR 4,080 crore (US $540 million) has been disbursed. We expect to disburse the rest within the next financial year. Spending on the hydropower projects has temporarily slowed because in two of the three projects, Punatsangchu-I and Punatsangchu-II, we have faced some geological surprises recently. There has been a slippage of river bank on one project which we did not anticipate when studies were made. We are trying to resolve this with the help of technical experts, including technical experts from abroad. In Punatsangchu-II, the surge gallery has had some problems from which we evacuate water.[37]

In July 2018, then Prime Minister of Bhutan Tshering Tobgay paid a visit to India. During his visit, he thanked the Government of India for its invaluable support and partnership for Bhutan's socio-economic development. During his meeting with Narendra Modi, the two leaders

> reviewed the bilateral economic and hydro-power co-operation, including the progress in implementation of the on-going [Government of India] GoI-assisted hydro-electric projects in Bhutan. They agreed that the robust partnership in the hydro-power sector was mutually beneficial. They reaffirmed their commitment to further strengthen collaboration in the hydro-power sector in the spirit of close friendship and co-operation that characterize relations between India and Bhutan.[38]

Soon after getting elected for a second term, in August 2019 Modi paid a visit to Bhutan where the two countries signed a number of Memorandum of Understanding (MoU) and other agreements. Modi and Lotay Tshering, Bhutan's Prime Minister, formally inaugurated the 720 MW Mangdechhu Hydroelectric Plant located at Trogsa Dzonkhag in Bhutan. With the coming into operation of this project, the jointly created hydropower generation capacity in Bhutan has crossed 2000 MW.[39] They expressed satisfaction on it and resolved to continue working together to expedite the completion of other ongoing projects such as Punatsangchhu-1, Punatsangchhu-2, and Kholongchhu HEP project.[40] India and Bhutan also reviewed ongoing bilateral discussions on the Sankosh Reservoir hydroelectric project. Bhutanese stamps commemorating five decades of mutually beneficial India-Bhutan cooperation in the hydropower sector was jointly released by Tshering and Modi.[41]

The larger quantity of the electricity generated by the Mangdechhu HEP project will meet the energy requirements of Bhutan and the surplus, around 3 billion units every year, will be exported to India.[42] Electricity will be transmitted to the Jigmeling substation in Bhutan by way of two 80-kilometres-long 400kV double-circuit transmission lines. Then from Jigmeling in Bhutan electricity will be supplied to Salakati in Assam.[43] The Power Purchase Agreement between Power trade Cooperation India Limited and Druk Green Power Corporation Limited for Sale and Purchase of Mangdechhu Power have been signed. This project is also expected to neutralise around 2.2 million tonnes of carbon dioxide from the atmosphere every year.[44] Earlier, in April 2019, the tariff protocol was signed between India and Bhutan which formalised the tariff at Indian Rs 2.4 (US $0.032 cents) per unit (kWh) for a period of 35 years.[45] This will be increased by 10% every five years until the loan is repaid and 5% thereafter.[46]

India's State owned National Hydroelectric Power Corporation (NHPC) acted as the design and engineering consultant for the Mangdechhu plant. Jaypee Group subsidiary Jaiprakash Associates won the contract for the construction of the dam, diversion tunnels, underground powerhouse, shafts, and the intake structures for the Mangdechhu project.[47] It subcontracted Bauer for the installation of seepage cut-off walls at cofferdams and related foundation engineering works, enabling the installation of the main dam. Bharat Heavy Electricals Limited got an Indian Rupees 5,940 million (US $786 million) contract to provide electromechanical equipment package for the project.[48] It includes the manufacture, supply, erection, and commissioning of four Pelton turbines and generators, control system, and other auxiliary equipment. Alstom T&D India is the supplier of power transformers for the project. The company supplied 13 units of 75MVA, 400 kilo volts generator transformers and an 80MVAR, 400kV shunt reactor, as part of the Indian Rupees 550 million (US $72.8 million) contract.[49] Bhutan Power Corporation built transmission lines from Mangdechhu plant to Jigmeling. Other contrators involved in this project were: PES Engineers, Kalpataru Power, Gammon India, Marti India, Encardio-rite, Central Electricity Authority (CEA), and Water and Power Consultancy Services (WAPCOS).[50]

To export electricity from Bhutan to India transmission grids are required. For it, the National Transmission Grid Master Plan (NTGMP) was prepared jointly by the erstwhile Department of Energy (DoE), now called Department of Hydropower & Power Systems (DHPS), Ministry of Economic Affairs, Royal Government of Bhutan and the Central Electricity Authority (CEA), Ministry of Power, Government of India between 2010–2012, at a cost of 18 million NU (US $2,38,178.5) under the Government of India's Power Trade Agreement financing in the Tenth Financial Year Plan.[51]

Concerns over India-assisted Hydroelectric Power Projects in Bhutan

Despite India's assistance to Bhutan in harnessing its HEP potential which contributes a maximum in its GDP, there are certain concerns among a section of Bhutanese over such HEP projects. Major reason for such concern is the increasing hydropower debt of Bhutan.

Close to 85% of Bhutan's trade is with India consisting mainly of HEP. On an average Bhutan has been supplying around 5,000–5500 million units every year to India.[52] India also export a limited unit to Bhutan, mainly during the dry season in Bhutan. In 2017, Bhutan imported electricity worth 74.94 million NU (US $0.99 million) from India and exported it electricity of worth 11,983.49 million NU (US $1585 million). This made trade balance in its favour of around 11,908.55 million NU (US $1575 million).[53] However, when other traded goods are included, the balance tilts in favour of India.

In 2018 primary polls though relationships with India and hydropower projects were not major issues, the two leading political parties – Druk Nyamrup Tshogpa (DNT) and Druk Phuensum Tshogpa (DPT) – mentioned both in their respective manifestos. On the relationships with India, the DPT's election manifesto stated:

> [The Party will] remain committed to maintaining and furthering the excellent relations with the people and the Government of India; carry forward the exemplary and mutually beneficial cooperation that is the hallmark of our [India-Bhutan] relations and deepen our economic ties.[54]

While the manifesto of the DNT, the top performer in the elections, did not specifically mention foreign policy and international diplomacy, it mentioned the development of hydro infrastructure in Bhutan with India's assistance. It also highlighted the development and use of reservoir-based hydropower to meet the peak electricity demand in India.[55] In the final elections which were held in October 2018 DNT won the elections and form the next government. In its manifesto, on HEP projects, the DNT mentioned "limited youth employment opportunities in the [hydropower] sector, especially in view of the nature of jobs".[56]

One of the complaints Bhutan has is that India buys cheap electricity from the hydroelectric projects in Bhutan. For example, in 2017, the tariff rate on the import of hydroelectricity from the Tala hydroelectric project by India was 1.80 Bhutanese Ngultrum (BTN) (US $0.24 cents) per unit. This was much below the domestic market price in India, which was around Indian Rupee 7 (US $0.093 cents) to 8 (0.11) per unit.[57] In July 2017, Bhutan proposed a tariff hike from the Chukha hydropower project. The last hike was made in 2014 when the tariff rate was raised from BTN 2 (US $0.26 cents) to BTN 2.25 (US $0.27 cents) per unit. The proposal to raise the tariff was then accepted, in principle, by India.[58]

Second, the guidelines issued by the Indian Cross Border Trade of Electricity (CBTE) in December 2016 are seen as unfavourable to Bhutan. Clause 5.2.1 of the guidelines states (mentioned in the introduction) is contrary to the Bhutan's Sustainable Hydropower Development Policy of 2008 which calls for having a minimum 51% shareholding by the Royal Government of Bhutan undertaking in any projects involving the other country.[59] In 2015–16 a few changes were made in the 2008 Hydropower Development Policy of Bhutan after a report by an ad hoc Committee on Hydropower Development Policy and Programmes.[60] Therefore the 2016 guidelines in the CBTE created confusion and developed certain apprehensions on the future of hydropower arrangements between India and Bhutan. In July 2017, in an effort to clarify the provisions of the CBTE guidelines, a Bhutanese delegation met Indian officials in New Delhi. During the meeting, the Indian officials gave the assurance that the guidelines would not apply to or impact the government-to-government projects between India and Bhutan. For non-government projects, as the Indian officials assured the Bhutanese delegation, there is a provision for a case-by-case clearance in the guidelines.[61]

In 2018 changes were made in the CBTE's 2016. Subsequent to it, works on the Kholongchhu project whose foundation stone was laid down during Modi's visit to Bhutan in 2014 was stopped and started again in 2018 after amendment in the guideline was made. The revised amendment has removed the earlier provision, which stated that only companies fully owned by the governments of the concerned countries or those having at least 51% equity investment of Indian public or private companies could export power to the Indian market after obtaining one-time approval from the designated authority in India.[62]

Dagachhu project, which was commissioned in 2015, also faced problems related to market access because of the CBTE guidelines.[63] In this project, Druk Green Power Corporation (DGPC) holds 59% equity, 26% is held by India's Tata Power Company Limited, and another 15% is held by the National Pension and Provident Fund.[64]

The new guideline in the amendment states that "the designated authority in India shall grant approval for export or import of electricity only after taking into account the generation capacity (as available) and the demand".[65] It means that imports to India may be permitted only when the demand exceeds generation capacity of electricity and exports may be permitted in case of extra availability.

The issue over the 1125 MW Kuri-I or Dorjilung project, supposedly a tripartite project between Bhutan, Bangladesh, and India, is not very clear now, Bhutan says that it is still studying the amendment. On tripartite agreement for transaction across India, the guideline says that

> Where tripartite agreement is signed for transaction across India, the participating entities shall sign transmission agreement with Central Transmission Utility of India for obtaining the transmission corridor access. Further the transmission system in India for transmission of electricity across the territory of India under cross border trade of electricity shall be built after concurrence from Government of India and necessary Regulatory approvals.[66]

Third, there are concerns in Bhutan over the delay in the completion of the HEP projects.

Delays in commissioning of the HEP projects are mainly due to geological surprises and other reasons that have direct impact on both growth and domestic revenue.[67] For example, the loan granted by India to Bhutan increases by 10%[68] every year which adds to the debt for the country.[69] Dasho Karma Ura, President of the Centre for Bhutan Studies and Gross National Happiness Research, stated, "It is important that hydropower, which is [a] key issue for the Bhutanese people also be looked at more quickly . . . public opinion in Bhutan was beginning to question the viability of the debt incurred by the projects".[70]

Finally, with the implementation of the Goods and Services Tax in India in 2017, its exports to Bhutan are cheaper than the imports from Bhutan.[71] This has an impact on Bhutan's trade deficit with India.

Besides, there are certain environmental related concerns. Former Economic Affairs Minister of Bhutan, Norbu Wangchuk said that as all projects are run-of-river, they are not harmful to environment.[72] At present, it is estimated that Bhutan offsets 4.4 million tons of carbon dioxide through exports of hydroelectricity this is expected to be around 22.4 million tons by 2025. Norbu claimed that the social and environmental cost of HEPs are low. The construction of a project involves only temporary damage.[73] These statements do not satisfy many environmentalists in Bhutan who are critical about the environmental impact of the hydropower. "Hydropower projects certainly harm the environment . . . both during and post construction", as Yeshey Dorji, an environmentalist and wildlife photographer in Bhutan observes.[74] Shripad Dharmadhikari, a coordinator of India-based non-profit group Manthan Adhyayan

Kendra, said that the project site for two Punatsangchhu projects in Bhutan was one of the habitats of the endangered White Bellied Heron. There are only about 200 such birds that remain globally.[75] According to Bhutanese media, the planned 540 MW Amochhu Reservoir Hydroelectric Project would likely displace the oldest indigenous community of Lhop or Doya people from their place. Dorji feels that the big problem is that in Bhutan hydropower projects are implemented secretly, some even without environmental clearances.[76] The reasons for such secrecy are the involvement of Indian companies in such projects. The Bhutanese officials do not want to upset India by creating any problems over sanction and construction of HEPs.[77]

The HEP projects are also seen as dangerous constructions because Bhutan lies on the Himalayan fault line which is responsible for the earthquake that devastated Nepal in 2015. It is predicted that the large earthquakes could hit the region in coming decades.[78] It has been noticed that such concerns hardly matter at the time of giving contract to build hydroelectricity projects in Bhutan. An example of it is both Punatsangchu I and II, which were awarded without taking into account such concerns. According to a joint study by the Comptroller and Auditor General India and Royal Audit Authority of Bhutan, the PI was awarded to Larsen and Tubro Company without investigating the extent of "geological surprises" it could cause.[79] Later, as some geological faults were found, works were stopped, and the project is now facing a delay from 2019 to 2022. Likewise, in January 2016 it was found that the Punatsangchhu-II comes across a shear zone in its downstream. As a result, this project has been delayed. Such delays also add financial burdens to Bhutan which is already under a huge hydro debt.[80] The Punatsangchhu-I dam was initially expected to cost BTN 40 billion (US $52.9 million), but it has increased to BTN 100 billion (US $13.23 billion) due to geological challenges and related delays.[81]

Sankosh HEP project is going to be a reservoir-based dam. It has a potential to generate 2,500 MW which is two times more than any of Bhutan's already operating hydropower projects. The Kuri-Gongri and Manas projects which are also reservoir-based would have more capacity than Sankosh. All three are on the southern border of Bhutan with India.[82] There is a proposal for a canal in the Sankosh project which would run through the Indian state of West Bengal into the Teesta River.[83] This proposed 60-metre-wide and 141-kilometres-long canal would cut right through the Buxa tiger reserve, which would affect the wildlife of the reserve. The canal may have an impact on Teesta River water flows also which is already an issue between West Bengal and Bangladesh. Punatsangchhu-I and Punatsangchhu-II plants are being built upstream of the Sankosh plant, creating nearly 5,000 MW from the same river system.[84] Once completed, the Sankosh project will be the most expensive hydropower project ever undertaken in Bhutan at BTN 115 billion (US $15.2 billion).[85]

Vasudha foundation study report

Many of such concerns call for an in-depth study of the India-Bhutan hydro projects. Unfortunately, only few such studies have been made. One of these, "A Study of the India-Bhutan Energy Cooperation Agreements and the Implementation of Hydropower Projects in Bhutan", was prepared by the New Delhi-based Non-Government Organisation (NGO), Vasudha Foundation in 2016.

According to the report, the first major challenge faced by Bhutan is hydropower project-related debts. Citing the findings of the Annual Report of the Royal Monetary Authority of Bhutan, the report stated that in 2014 Bhutan's Indian Rupee debt constituted 64% of Bhutan's total debt and HEP projects related loans accounted for 83.4% of the total Rupee loan. The actual interest payments on Rupee-denominated HEP debt amounted to Indian Rupees 1.4 billion (US $1.85 million) in 2014 and accrued interest on the three ongoing hydropower projects (Punatsangchhu-I, Punatsangchhu-II and Mangdechhu (now completed)) amounted to almost Indian Rupees 3.6 billion (US $4.76 million).[86] This has increased in subsequent years. For example, in the financial years 2016–17, the country had Indian Rupees 118.77 billion (US $156.77 million) of outstanding Indian Rupees loans. Of the total Rupee debt, 94.11% was public debt on HEP while 5.89% represented debt taken to finance BOP [Balance of Payments] transactions with India (the Government of India line of credit).[87]

It is being maintained that the three projects – Mangedchhu (now completed) and Punatsangchhu-I and II – would put Bhutan under about Indian Rupees 12,300 crore (US $162.9 billion) rupees of debt. This accounts for 77.52% of the country's total debt and is about 72.95% of its GDP[88] which has increased while the non-hydro debts have been decreased to 4.02 billion NU (US $5.32 million), i.e., 20.21% of the Bhutan's GDP.[89]

According to the *Annual Financial Statement*, 2018 of Bhutan, of the total external debt, hydropower debt constituted 132,532.919 million NU (US $1753.69 million) accounting for 74.43% of the total external debt while the non-hydropower debt stock was 44,619.477 million NU (US $590.04 million) accounting for 25.57% of the country's GDP. Loans from India were Indian Rupees 133,190.701 million (US $17.64 billion) with Indian Rupees 119,452.841 million (US $15.82 billion) as HEP debt stock representing 89.69% of the Bhutan's outstanding debt to India.[90]

Second, with an increase in engagements of the Indian private companies in the hydropower sector, and the initiation of the joint venture projects, the Bhutanese feel that they are not equal participants in exploring their water resources for their own benefits. It has been reported that there is only a marginal participation of the Bhutanese private sector in HEP. Normally, they act as small subcontractors in civil works such as supplying boulders and sands. According to the Bhutan Chamber of Commerce & Industry

(BCCI), the Bhutanese private sector has not reaped the benefits of hydro-power development in the country. The BCCI has pushed for a greater role to local companies in hydropower construction. In 2013, the BCCI argued for the transfer of those works to the Bhutanese companies which they could do instead of inviting the Indian companies to do them.[91]

Third, apart from the HEP projects, there is also a concern over the emergence of an auxiliary Indian economy in Bhutan. The massive influence of Indian hydropower companies and their participation in the accompanying sectors are being questioned by the Bhutanese. For example, Larsen & Tubro, Gammon India, and Hindustan Construction Companies have also set up stone crushing industries, besides their contract in the Punatsangchhu hydropower project.[92]

Fourth, Vasudha report has talked about relations between ecology and HEP. In Bhutan, the Department of Forest and Park Services looks after the ecology of the country. As a norm, any HEP project has to clear the Environmental Impact Assessment (EIA) test before construction works can start. In Bhutan, the EIA was not carried out for projects whose construction began before 2000. Therefore, the environmental impact study of such HEP project s as Chhukha, Kurichhu and Tala was not conducted. The project authorities for Punatsangchhu-I, Punatsangchhu-II, and Mangdechhu conducted the EIA but the report has not been made public.[93] Then, as mentioned earlier, HEP projects have been warned for throwing "geological surprises".[94]

Fifth, many people from the project-affected areas have claimed that their consent was not sought for such hydropower projects as Tala, Punatsangchhu-I, Punatsangchhu-II, and Mangdechhu although the district authorities shared information about these projects with them. During the information sharing session, the adverse impact of these projects was not discussed with the people. Under the Land Act of 2007, whenever land is acquired by the authorities to develop hydropower project, the land holder has the choice of opting for either land or money as compensation.[95] The replacement land should be of equal quality as the land taken by the government for project development. However, in many cases, the villagers were given barren land as compensation against their agricultural land.[96]

Sixth, the hydropower projects have also affected the natural flow of water in ponds and/or spring water. Most of the sources have been diverted to channel the reservoir or tunnel for run-of-the-river multipurpose projects. The construction of Mangdechhu I has also disrupted waters supply through pipelines to the people.[97] Bhutan does not have a proper resettlement and rehabilitation policy for the people affected by the hydropower projects. Under the Hydropower Policy of 2008, the project authorities have to set aside a minimum 1% of the project cost for the resettlement and rehabilitation of affected people. In most cases, there is misuse due to the absence of a monitoring body.[98]

Seventh, according to the Vasudha Foundation report, the HEP projects have not provided enough employment opportunities to the Bhutanese. According to a release by the Indian Embassy in Bhutan in 2015, "There are about 60,000 Indian nationals living in Bhutan, employed mostly in the hydro-electric power and construction industry. In addition, between 8,000 and 10,000 daily workers enter and exit Bhutan every day in [the] border towns".[99] The rate of unemployment in Bhutan was 2.1% in 2013 and 2.9% in 2014.[100] At many project sites and after the start of their operations, some of the locals allege that they have not been provided with the necessary training to take up jobs in the sector.[101] This is being contradicted by the Royal Monetary Authority Report of 2016–17 which says that the hydropower sector has contributed to generating direct and indirect employment opportunities for the Bhutanese. The report states,

> As of June 2017, there are 3,950 (851 are key officials and staff) employees in PHP I (Punatsangchu Project-I), 6,942 (857 key officials and staff) in PHP II (Punatsangchu Project-II), and 4,857 (509 key officials and staff) in MHP (Mangdechhu Project). There are about 1,687 employees under DGPC (Druk Green Power Corporation), mainly for operation and maintenance of generating hydro plants, and about 2,500 under Bhutan Power Corporation for the construction and maintenance of electricity transmission line in Bhutan.[102]

Further, the 11th planning commission report (2013–2018) mentions that in 2013 the number of employees in the country's hydropower sector was 5853; this was increased to 12,727 by 2017. It means that in the five years' time around 6,874 jobs were created in hydroelectricity sector.[103]

On the findings of the Vasudha foundation's report, then Bhutanese Prime Minister Tshering Tobgay said that "If they are concerned about our environment, I am also concerned. They should have met concerned people from the government before publishing".[104] He told *thethirdpole.net* that "I agree hydropower projects do damage the environment. But the hydropower projects in Bhutan, which are run-of the river schemes, do the least damage".[105] He added, "When we have hydropower, we do not have to use thermal power, or power generated from nuclear sources or coal. That is good for the environment. In that context hydropower is good for environment".[106]

Then, speaking to *thethirdpole.net*, the managing director of Druk Green Power Corporation, Chhewang Rinzin, said

> Yes, I suppose we could improve on what we are doing, but this would require more than the hydropower sector which only provides the money to rehabilitate and resettle. It is for the other agencies to actually implement it. This would require changes to the Land Act, Environment Act, Forestry Act, the power rate, the working of the Ministry of Home and Cultural Affairs, and many other areas.[107]

He also added

> In many ways the [Vasudha] report helps us in understanding that maybe we need to do little bit more, but it would have been better if the report had mentioned whether we are following the laws and policies of the Kingdom of Bhutan. . . . But I do agree that, even without Vasudha Foundation's report pointing it out, maybe we need to do a little bit more. Today many bigger projects that started have come to stop because of political reasons. . . . From that point of view, I think Bhutan has to learn and ensure that we do not get into same type of trap. That would have been more useful than the existing report, which was not balanced, and seemed to accuse us of doing nothing. We are investing millions and millions into environment and social impact assessment and also into implementing the mitigating measures, and this was not represented [in the report].[108]

Conclusion

Bhutan is a close partner of India. The latter's grants and loans to develop the hydropower sector are crucial to Bhutan's economy. However, in recent years, with changes in India's policy on constructing HEP projects in the neighbourhood, the grant amounts to Bhutan have decreased. There is a prevailing undercurrent in Bhutan that India is exploring and exploiting Bhutan's water sector for its own benefit. This is part of the reason why Bhutan has refused to sign the concession agreement for some of the joint venture projects.[109]

Besides, a number of concerns and questions a section of Bhutanese have against the India's HEP projects have to be properly and timely addressed by India. Many of such concerns and unanswered questions create doubts in their minds, due to which often projects get delayed and unnecessarily add to existing hydro debt of Bhutan.

As mentioned in Chapter 1, with growing China's military build-up in Bhutanese territory near the Indian border, it is extremely important for India to have a friendly Bhutan. In 2017 and even earlier during an India-China military stand-off at Doklam Bhutan had extended its support to India. To maintain a similar rapport with Bhutan, it is necessary for India to continue to provide concessionary grants to the country. To India's dismay, in Bhutan the number of people talking about opening up to China and establishing diplomatic relationships with it is growing.[110]

Notes

1 Food an Agricultural Organisation, 'Survey of Waters of Bhutan Physiological' http://www.fao.org/3/L8853E/L8853E02.htm. Accessed on 10 June 2018.
2 "Glacial Lake Outburst Flood". *Report of the International Conference: Reducing Risks Ensuring Preparedness.* http://www.undp.org/content/dam/undp/

documents/projects/BTN/Proceeding%20of%20International%20GLOF%20 Conference%20_Dec%202013%20Paro.pdf.

3 Shailendra, Yashwant. (2018, 27 August). "Villagers in Bhutan and India Come Together to Share River". *Thethirdpole.net*. www.thethirdpole.net/en/ 2018/08/27/villagers-in-bhutan-and-india-come-together-to-share-river/. Accessed on 27 September 2018.

4 Asian Development Bank. (2016). "Water Securing Bhutan's Future". www.adb. org/sites/default/files/publication/190540/water-bhutan-future.pdf. Accessed on 2 April 2018, p. 75.

5 Royal Government of Bhutan, National Environment Commission. (2014). "Bhutan Water Vision and Bhutan Water Policy". www.nec.gov.bt/nec1/wp-con tent/uploads/2014/04/Bhutan-Water-Policy-Eng.pdf. Accessed on 2 April 2018.

6 Ibid., p. 12.

7 Royal Government of Bhutan, National Environment Commission. "National Integrated Water Resources Management Plan 2016". www.nec.gov.bt/nec1/wp-content/uploads/2016/03/Draft-Final-NIWRMP.pdf. Accessed on 21 May 2018, p. 33.

8 Ibid.

9 Walker, Beth. (2015, 20 May). "Will Mega Dams Turn Bhutan's Happiness Sour?" *The Guardian*. www.theguardian.com/sustainable-business/2015/may/20/will-mega-dams-turn-bhutans-happiness-sour. Accessed on 4 December 2018.

10 Royal Government of Bhutan, National Environment Commission. "National Integrated Water Resources Management Plan 2016". www.nec.gov.bt/nec1/wp-content/uploads/2016/03/Draft-Final-NIWRMP.pdf. Accessed on 21 May 2018.

11 Royal Government of Bhutan, National Environment Commission. (2014). "Bhutan Water Vision and Bhutan Water Policy". www.nec.gov.bt/nec1/wp-content/uploads/2014/04/Bhutan-Water-Policy-Eng.pdf, p. 11.

12 Asian Development Bank. (2016). "Water Securing Bhutan's Future". www.adb. org/sites/default/files/publication/190540/water-bhutan-future.pdf. Accessed on 2 April 2018.

13 Dhrmadhikary, Shripad. (2016, 4 October). "India Bhutan Hydropower Cooperation Fraying at the Edges", *Thethirdpole.net*. www.thethirdpole. net/2016/10/04/india-bhutan-hydropower-cooperation-fraying-at-the-edges/. Accessed on 28 March 2018.

14 Druk Green Power Corporation Limited, Government of Bhutan. www.druk green.bt/index.php/about-us. & Status of Hydropower Dams in Bhutan. *International Rivers*, 14 April 2015. www.internationalrivers.org/resources/8703. Department of Renewable Energy, Ministry of Economic Affairs, Government of Bhutan. www.moea.gov.bt/?page_id=1140.

15 Planning Commission Royal Government of Bhutan. "Bhutan 2020: A Vision for Peace, Prosperity and Happiness, Part II". www.greengrowthknowledge. org/sites/default/files/downloads/policy-database/Bhutan%20Vision%20 2020%20II.pdf, p. 72.

16 Gross National Happiness Commission, Royal Government of Bhutan. (2018– 2023). *Twelfth Five Year Plan*. www.gnhc.gov.bt/en/wp-content/uploads/ 2019/05/TWELVE-FIVE-YEAR-WEB-VERSION.pdf. Accessed on 30 January 2020.

17 Ibid.

18 Ibid.

19 Arora, Vishal and Chencho Dema. (2016, 16 February). "Bhutan Should Come Clean on Hydropower Megaplan". *The Diplomat*. https://thediplomat. com/2016/02/bhutan-should-come-clean-on-hydropower-megaplan/. Accessed on 17 March 2018.

20 Ibid.

21 After its origin in the Indian state of Sikkim, Jaldhaka flows in India, Bhutan, and Bangladesh.

22 Ghosh, Shubham. (2014, 18 June) "Understanding India-Bhutan Relations". *One India*. www.oneindia.com/feature/understanding-india-bhutan-relations-1467521.html. Accessed on 27 March 2018.

23 Ministry of Foreign Affairs, Royal Government of Bhutan. "Bhutan-India Hydropower Relations". New Delhi: Royal Bhutanese Embassy. www.mfa.gov. bt/ rbedelhi/?page_id=28. Accessed on 16 June 2018.

24 "Agreement Between the Government of the Republic of India and the Royal Government of Bhutan Concerning Cooperation in the Field of Hydrolectric Power". www.internationalrivers.org/sites/default/files/attached-files/india_bhu tan_hydropower_agreement_july_2006.pdf.

25 Ministry of External Affairs, Government of India. "India-Bhutan Relations". https://www.mea.gov.in/Portal/ForeignRelation/Bhutan_May_2018.pdf. Accessed on 16 September 2016.

26 Protocol to the 2006 Agreement between the Government of Republic of India and the Royal Government of Bhutan concerning cooperation in the field hydroelectric power. www.internationalrivers.org/sites/default/files/attached-files/india_ bhutan_hydropower_protocol_march_2009.pdf.

27 Ibid.

28 "Tala Hydroelectric Project". www.power-technology.com/projects/tala/. Accessed on 20 April 2018.

29 Ibid.

30 Committee on External Affairs (2017–18) Sixteenth Lok Sabha, Ministry of External Affairs, Demand for Grants (2018–19) Twenty First Report, Lok Sabha Secretariat. http://164.100.47.193/lsscommittee/External %20Affairs/16_Exter nal_Affairs_21.pdf. Accessed on 22 March 2018, pp. 70–71.

31 Ibid.

32 Ibid.

33 Unlike the run-of-the river projects, reservoirs store large quantity of waters. Run-of-the-river projects have pondage to store limited amount of waters.

34 Druk Green Power Corporation Limited, Government of Bhutan. www.druk green.bt/index.php/about-us. & Status of Hydropower Dams in Bhutan *International Rivers* 14 April 2015. www.internationalrivers.org/resources/8703. Department of Renewable Energy, Ministry of Economic Affairs, Government of Bhutan www.moea.gov.bt/?page_id=1140.

35 Ibid.

36 Kholongchu Hydro Energy Limited. www.khepbhutan.com/. Accessed on 12 June 2018.

37 Committee on External Affairs (2017–18) Sixteenth Lok Sabha, Ministry of External Affairs, Demand for Grants (2018–19) Twenty First Report, Lok Sabha Secretariat. http://164.100.47.193/lsscommittee/External %20Affairs/16_Exter nal_Affairs_21.pdf. Accessed on 22 March 2018, p. 73.

38 Ministry of External Affairs, Government of India. "Official Visit of the Prime Minister of Bhutan to India". www.mea.gov.in/press-releases.htm?dtl/30036/ official+visit+of+the+prime+minister+of+bhutan+to+india. Accessed on 12 July 2018.

39 Ministry of External Affairs, Government of India. "Joint Statement on the State Visit of Prime Minister of India to Bhutan". www.mea.gov.in/bilateral-documents.htm?dtl/31739/Joint_Statement_on_the_State_Visit_of_Prime_Min ister_of_India_to_Bhutan. Accessed on 1 September 2019.

40 Ibid.

41 Ibid.

42 Sassi, Anil. (2019, 20 August). "Tale of Two NHPC Projects: Bhutan on, India off". *The Indian Express*. https://indianexpress.com/article/india/india-bhutan-nhpc-projects-hydroelectric-power-plant-5918599/. Accessed on 22 August 2019.
43 Ibid.
44 "Mangdechhu Hydroelectric Project, Bhutan". *Power Technology*. www.power-technology.com/projects/mangdechhu-hydroelectric-project-trongsa-dzong khag/. Accessed on 30 January 2020.
45 Sassi, Anil. (2019, 20 August). "Tale of Two NHPC Projects: Bhutan on, India off". *The Indian Express*. https://indianexpress.com/article/india/india-bhutan-nhpc-projects-hydroelectric-power-plant-5918599/. Accessed on 22 August 2019.
46 Ibid.
47 "Mangdechhu Hydroelectric Project, Bhutan". *Power Technology*. www.power-technology.com/projects/mangdechhu-hydroelectric-project-trongsa-dzong khag/. Accessed on 30 January 2020.
48 Ibid.
49 Ibid.
50 Ibid.
51 Department of Hydropower & Power Systems, Ministry of Economic Affairs, Royal Government of Bhutan National Transmission Grid Master Plan (NTGMP) of Bhutan-2018. www.moea.gov.bt/wp-content/uploads/2018/11/National-Transmission-Grid-Master-Plan-2018.pdf.
52 Press Information Bureau, Government of India, Ministry of Power. (2017, 29 March). "India Becomes Net Exporter of Electricity for the First Time". http://pib.nic.in/newsite/PrintRelease.aspx?relid=160105.
53 Department of Renewable Energy, Ministry of Economic Affairs, Government of Bhutan. "Annual Trade Statistics, 2017". www.moea.gov.bt/wp-content/uploads/2017/10/ANNUAL-TRADE-STATISTICS-2017.pdf, pp. 4–5.
54 Druk Phuensum Tshogpa Manifesto. (2018). "In Pursuit of Gross National Happiness: Equity and Justice". www.ecb.bt/pp/dpt/dptmanifesto2018.pdf. Accessed on 22 September 2018.
55 Ibid.
56 Druk Nyamrup Tshogpa Manifesto. (2018). "Narrowing the Gap". file:///C:/Users/isasar/Downloads/DNT-Manifesto-2018-1%20(2).pdf. Accessed on 22 September 2018, pp. 22–30.
57 Tenzing, Lamsang. (2018, 26 July). "More Than Doklam Issue, Bhutan Worries About Hydropower Deficits". *The Indian Express*. http://indianexpress.com/article/opinion/more-than-the-doklam-issue-bhutanworried-about-hydropower-deficits-4768598/. Accessed on 27 July 2017.
58 "Bhutan to Supply Hydropower to Bangladesh Via India". *Outlook*, 2017, July. www.outlook india.com/newsscroll/bhutan-to-supply-hydropower-to-bangla desh-via-india/1090541. Accessed on 19 June 2018.
59 Ministry of Economic Affairs. "Bhutan Sustainable Hydropower Development Policy, 2008". www.moea.gov.bt/wp-content/uploads/2017/07/Hydropower-Policy.pdf. Accessed on 17 October 2018.
60 See "Report to the 16th Session of the National Council on Hydropower Development Policy and Programmes by Ad hoc Committee, 2015". www.nation alcouncil.bt/assets/uploads/files/Hydro%20Report%20as%20on%20Nov%20 26%202015-%20Final%20for%20deliberation%202.pdf. Accessed on 17 October 2018.
61 "Bhutan to Supply Hydropower to Bangladesh Via India". *Outlook*, 2017, 2 July. www.outlook india.com/newsscroll/bhutan-to-supply-hydro power-to-bangladesh-via-india/1090541. Accessed on 19 June 2018.

62 Dorji, Tshering. (2018, 25 December). "India Amends Regulations on Cross-Border Electricity Trade". *Kuensel*. www.kuenselonline.com/india-amends-regulations-on-cross-border-electricity-trade/. Accessed on 12 July 2019.

63 Ibid.

64 Ibid.

65 Ibid.

66 Government of India, Ministry of Power. "Guidelines for Import/Export (Cross Border) of Electricity-2018- Regarding". Office Memorandum. https://powermin.nic.in/sites/default/files/uploads/Guidelines_for_ImportExport_Cross%20Border_of_Electricity_2018.pdf. Accessed on 25 January 2019.

67 Gross National Happiness Commission, Royal Government of Bhutan. (2018–2023). *Twelfth Five Year Plan*. www.gnhc.gov.bt/en/wp-content/uploads/2019/05/TWELVE-FIVE-YEAR-WEB-VERSION.pdf. Accessed on 30 January 2020.

68 At some places it is mentioned that the rate is 9%. See "Tala Hydroelectric Project". www.power-technology.com/projects/tala//. Accessed on 20 April 2018.

69 Haider, Suhasini. (2017, 6 September). "Hydropower Debt, Delays Biggest Challenge in Ties with India, Say Bhutan Officials". *The Hindu*. www.thehindu.com/news/national/hydropower-debt-delays-biggest-challenge-in-ties-with-india-say-bhutan-officials/article19630701.ece. Accessed on 2 April 2018.

70 Ibid.

71 Tenzing, Lamsang. (2018, 26 July). "More Than Doklam Issue, Bhutan Worries About Hydropower Deficits". *The Indian Express*. http://indianexpress.com/article/opinion/more-than-the-doklam-issue-bhutanworried-about-hydropower-deficits-4768598/. Accessed on 27 July 2017.

72 Arora, Vishal and Chencho Dema. (2016, 16 February). "Bhutan Should Come Clean on Hydropower Megaplan". *The Diplomat*. https://thediplomat.com/2016/02/bhutan-should-come-clean-on-hydropower-megaplan/. Accessed on 17 March 2018.

73 Ibid.

74 Ibid.

75 Ibid.

76 Ibid.

77 Ibid.

78 Walker, Beth. (2015, 20 May). "Will Mega Dams Turn Bhutan's Happiness Sour?" *The Guardian*. www.theguardian.com/sustainable-business/2015/may/20/will-mega-dams-turn-bhutans-happiness-sour. Accessed on 4 December 2018.

79 Ahmad, Omair. (2015, 30 July). "Losing the Dragon – India-Bhutan Relations One Year After Modi's Historic Visit". *The Hindu*. www.thehindu.com/opinion/op-ed/prime-minister-narendra-modis-historic-visit-to-bhutan/article7480714.ece. Accessed on 24 July 2018.

80 Ibid.

81 Ahmad, Omair. (2019, 24 June). "Many Questions as Bhutan Plans Big Dam with Reservoir". *Thethirdpole.net*. www.thethirdpole.net/en/2019/06/24/bhutan-prioritises-reservoir-hydropower/. Accessed on 27 June 2019.

82 Ibid.

83 Ibid.

84 Ibid.

85 Ibid.

86 Vasudha Foundation. (2016, January). "A Study of the India-Bhutan Energy Cooperation Agreements and the Implementation of Hydropower Projects in Bhutan". Vasudha Foundation. www.vasudha-foundation.org/wp-content/uploads/Final-Bhutan-Report_30th-Mar-2016.pdf. Accessed on 28 March 2018, p. 28.

87 "Royal Monetary Authority of Bhutan, Government of Bhutan, Annual Report 2016–17". www.rma. org.bt/RMA%20Publication/Annual%20Report/ annual%20report%20%202016-2017.pdf. Accessed on 28 March 2018, p. 30.

88 Haider, Suhasini. (2017, 6 September). "Hydropower Debt, Delays Biggest Challenge in Ties with India, Say Bhutan Officials". *The Hindu*. www.thehindu. com/news/national/hydropower-debt-delays-biggest-challenge-in-ties-with-in dia-say-bhutan-officials/article19630701.ece. Accessed on 2 April 2018.

89 Eleventh Five Year Plan 2013–2018, Gross National Happiness Commission Royal Government of Bhutan. (2018, June). www.gnhc.gov.bt/en/wp-content/ uploads/2018/06/small1.compressed.pdf, p. 3.

90 Ministry of Finance, Royal Government of Bhutan. "Annual Financial Statement 2018". www.mof.gov.bt/wp-content/uploads/2019/05/AFS2017-18Eng lish.pdf, p. 30.

91 Vasudha Foundation. (2016, January). "A Study of the India-Bhutan Energy Cooperation Agreements and the Implementation of Hydropower Projects in Bhutan". Vasudha Foundation. www.vasudha-foundation.org/wp-content/uploads/ Final-Bhutan-Report_30th-Mar-2016.pdf. Accessed on 28 March 2018, p. 27.

92 "The Rupee Crunch and India-Bhutan Economic Engagement". *IDSA, Medha Bisht*, 2012, 16 July. https://idsa.in/issuebrief/TheRupeeCrunchandIndiaBhu tanEconomicEngagement_MedhaBisht_160712. Accessed on 3 April 2018.

93 Vasudha Foundation. (2016, January). "A Study of the India-Bhutan Energy Cooperation Agreements and the Implementation of Hydropower Projects in Bhutan", p. 31. Vasudha Foundation. www.vasudha-foundation.org/wp-con tent/uploads/Final-Bhutan-Report_30th-Mar-2016.pdf. Accessed on 28 March 2018.

94 Ranjan, Amit. (2019, 8 June). "During Visit to Bhutan, Jaishankar Must Address Hydropower Issues". https://thewire.in/diplomacy/india-bhutan-jais hankar-visit-hydropower-projects. Accessed on 12 August 2019.

95 Vasudha Foundation. (2016, January). "A Study of the India-Bhutan Energy Cooperation Agreements and the Implementation of Hydropower Projects in Bhutan". Vasudha Foundation. www.vasudha-foundation.org/wp-con tent/uploads/Final-Bhutan-Report_30th-Mar-2016.pdf. Accessed on 28 March 2018, p. 37.

96 Ibid.

97 Ibid.

98 Ibid., pp. 38–39.

99 Ibid., p. 40.

100 Ibid.

101 Ibid.

102 "Royal Monetary Authority of Bhutan, Government of Bhutan, Annual Report 2016–17". www.rma. org.bt/RMA%20Publication/Annual%20Report/ annual%20report%20%202016-2017.pdf. Accessed on 28 March 2018, p. 50.

103 Eleventh Five Year Plan 2013–2018, Gross National Happiness Commission Royal Government of Bhutan. (2018, June). www.gnhc.gov.bt/en/wp-con tent/uploads/2018/06/small1.compressed.pdf, p. 101.

104 Walker, Beth. (2016, 4 October). "Bhutan's PM Defends Hydropower Dams Against Blistering Report". *Thethirdpole.net*. www.thethirdpole.net/en/2016/10/04/ bhutans-pm-defends-hydropower-dams-against-blistering-report/.

105 Ibid.

106 Ibid.

107 Gylmo, Dawa. (2016, 1 November). "The Future of Bhutan's Hydropower". *Thethirdpole.net*. www.thethirdpole.net/en/2016/11/01/the-future-of-bhutans-hydropower/.

108 Ibid.
109 Tenzing, Lamsang. (2018, 26 July). "More Than Doklam Issue, Bhutan Worries About Hydropower Deficits". *The Indian Express.* http://indianexpress.com/article/opinion/more-than-the-doklam-issue-bhutanworried-about-hydro power-deficits-4768598/. Accessed on 27 July 2017.
110 Ganpati, Nirmala. (2017, 28 July). "China-India Stand-off Has to Do With Bhutan". *The Straits Times.* https://www.straitstimes.com/asia/south-asia/china-india-stand-off-has-to-do-with-bhutan. Accessed on 1 August 2017.

Conclusion

As the population is increasing and the phenomenon of climate change has accelerated, the demand-supply gap of water in India and its South Asian riparian States have widened. Such a widening of the gap in water availability is leading to competition, contests, tensions, and disputes over shared rivers.

Some of these water-related disputes may further aggravate political tensions between the South Asian riparian States such as between India and Pakistan who do not share a long period of good political relationships since the Partition of British India in 1947. Growing demands of water in the border region may also disturb India's relationships with Bangladesh. Unlike Pakistan and Bangladesh, the upper riparian States, Bhutan, and Nepal may be asked or requested or, as some Nepali told me, "pressurised" by India to release more water to meet India's growing water demand. Nepal is being also held responsible for annual floods in the Indian state of Bihar. Higher precipitation in Nepal, due to the phenomenon of climate change, means the release of more waters, which may lead to more severe floods in Bihar.

In the beginning of the book, one of the arguments I have made is that the countries sharing good political relationships have agreed to make arrangements and have signed treaties to share transboundary rivers waters. Even some of those riparian States which do not share good relationships have agreements, Memorandum of Understanding (MoUs), or treaties to share common waters. They have done it bilaterally or after mediation by international organisation. In both cases the water sharing arrangements have not worked well for a long period, as voices have been raised against them due to rise in demand for waters. In South Asia, India shares good relationships with Bhutan, Nepal, and Bangladesh. It has transboundary water-related treaties with Nepal and Bangladesh and is engaged in Hydroelectric Power (HEP) projects development in Bhutan. Nevertheless, problems, differences, and disputes over them remain.

Although India and Bangladesh have signed a treaty on sharing Ganges waters, the two countries have issues on many other of the 54 transboundary rivers they share. Many of these rivers are border rivers and mark the boundary between India and Bangladesh. In 1947 India's border with East

Pakistan were also marked in the middle of these border rivers which still creates confusion. The emergence, re-emergence, and submersion of *chars* (sand silted lands) in the middle of rivers create confusions on the river borders between India and Bangladesh. Border rivers are also being used, as India accuses, for cattle smuggling, "illegal" immigration, and for other activities from Bangladesh side into Indian territory. To stop such activities, India has planned to do "smart fencing" along India-Bangladesh riverine border in Assam.[1]

On River Teesta waters, India and Bangladesh have differences and not serious disputes in technical term. This is so because the Union government of India is ready to release the percentage of waters agreed in the interim agreement of 2011, but the West Bengal government has reservations against it. Since 2011, Bangladesh has been unsuccessfully trying to persuade Mamata Banerjee to release the amount of waters agreed in the interim agreement. During her visit to India in 2017, Sheikh Hasina reiterated her demand to implement the interim agreement on Teesta water issue but failed to impress Mamata Banerjee. In May 2018, during her visit to West Bengal to receive an honorary Doctor of Literature degree conferred upon her by Kazi Nazrul University and attend the 49th Convocation at Viswa Bharti University, near Kolkata, Sheikh Hasina met Narendra Modi and Mamata Banerjee, Chief Minister of West Bengal. The three leaders did not talk about Teesta during their meetings.[2] Following a meeting with Hasina on 26 May 2018, when asked by journalists if there was any talk on the Teesta water sharing, Banerjee did not reply.[3]

India and Bangladesh are also looking for a "mutually acceptable formula" on water sharing from all 54 transboundary rivers.[4] During his visit to Dhaka from 19 to 21 August 2019 India's External Affairs Minister, S. Jaishankar, said that "We are [India and Bangladesh] ready to start from anywhere. . . . We look forward to making progress in finding a mutually acceptable formula to share water from 54 shared rivers".[5] Foreign Secretary-level meeting was held in August 2019 where it was decided to work on water sharing and basin management of seven rivers – the Manu, Muhuri, Khowai, Feni, Gumti, Dharla, and the Dudhkumar.[6] In the meeting, the two countries also decided to carry out a feasibility study on the proposed Ganges Barrage project, mainly its impact on the environment and people's livelihoods in the two countries.[7]

Like Bangladesh, India also shares good relationships with Nepal. The two countries are considered to be close. However, in recent times, their political relationships have been affected since India carried out an economic blockade against Nepal from September 2015 to February 2016, though the Indian government never accepts this. Afterwards, China has made strong inroads in Nepal. There has been a spike in trade between China and Nepal, though it favours the former over the latter. To counter China's footprint in Nepal, India has come up with several planned projects to help Nepal but their efficacy to change the growing anti-India perceptions among a section of Nepali has to be observed.

Even in the past, a section of Nepalis had problems with what they see as India's over interference in Nepal's affairs. As discussed in Chapter 1 of this book, a section of Nepalis have raised questions over the India-Nepal friendship treaty of 1950 since it was signed. On water issues, as discussed in Chapter 4, since the first water sharing treaty was signed in the 1950s, there has been opposition in Nepal. As a result, the Kosi (1954) and Gandaki (1959) agreements were amended in a few years' time after they were signed. In 1996 Mahakali Treaty was opposed by the parliamentarians of Nepal. On the ground, there has been also opposition to Indian hydropower projects in Nepal. They are looked at as one who exploits their waters for the benefits of India. In 2016, the provisions made by the CBTE has exploded the situation but amendments in 2018 have given relief to the Nepalis and helped to restart the works many of the hydroelectricity. Strong protests against Karnali and Kosi high dam projects have delayed the work.

With Bhutan, India is very close. The two countries do not have disputes over water sharing per se but there are issues over India's hydropower projects in Bhutan. India provides funds and technologies for almost all hydroelectricity projects (HEPs) in Bhutan. This has pushed Bhutan into huge hydro debt. Besides, there are environment and ecology related concerns of Bhutanese which rarely affects the hydropower sector of the country while making decisions about the construction of dams and hydroelectricity projects. Like Nepal, Bhutan also expressed concerns over CBTE's guidelines of 2016. Works on projects such as Kholongchhu were stopped. However, after amendments in the law, works have been started. Report by a Delhi-based Non-Government Organisation, Vasudha foundation, published in 2016, analysed in Chapter 4, has surfaced many concerns and issues related to India's HEPs in Bhutan.

Like Bangladesh, Bhutan, and Nepal, India does not share even "normal" relationships with Pakistan. The two countries have fought wars and remain in permanent state of tensions with little period of calculated bonhomie in past. In 2019 they have severed whatever little links they had with limited trade between the people from two countries. The High Commissioner designate of Pakistan was held back from assuming his office at New Delhi and Indian High Commissioner was sent back from Islamabad in 2019. India has even warned about the impact of such relations on the transboundary waters.

Notably, in 1960 when India and Pakistan signed the Indus Waters Treaty (IWT), they were not in a very good relationship. The IWT was largely possible because of the mediation from the World Bank. Soon after the IWT was signed the leaders were abhorred and criticised by the members of opposition and public in their respective countries. Even today such abhorrence continues. The IWT has worked throughout all tensions between the two countries, though voices have been made to scrap it. After the involvement of a Pakistan-based militant group in an attack on India's paramilitary force convoy in 2019, political statements from the Indian Prime Minister and the Water Resources Minister have raised a new debate on the future of the IWT.

Most of the India-Pakistan issues on the hydroelectricity projects on their shared rivers are political. In the past when the two countries were in a working relationship, they have agreed to resolve their differences bilaterally. For example, in 2010, and in recent times in 2019, as mentioned in Chapter 2, India and Pakistan shared data and information on some of their HEP projects on their shared rivers.

At present, the two governments are not on talking terms. This affects their water relationships. India has already decided to stop its share of waters from the eastern rivers from flowing into Pakistan. This is in accordance to the provisions of the IWT. However, it may open the floodgates, questioning the IWT's future.

Looking at the link between political relationships and the water-related disputes, one finds that, theoretically, as neo-realists would find, all riparian States behave in a rational way to secure maximum benefits from the water-related treaties and look for keeping most of the waters of transboundary rivers under its control. This is why despite having good relationships with Bangladesh, West Bengal gives precedence to its own water interests rather than making adjustments and sharing its water to further consolidate bilateral relations between India and Bangladesh. In the case of Nepal, even though India and Nepal share the same religious majority, as states they behave differently and look for secure its own interests. Almost all water-related agreement between India and Nepal and HEP projects assisted by India have faced resistance in Nepal. The critics see those agreements and projects favourable to India with less benefits to Nepal. Like Nepal, Bhutan is also a close friend of India. It also has issues with India's HEP projects which has pushed the country into HEP projects linked debt. Unlike Bangladesh, Nepal, and Bhutan, Pakistan does not share good relations with India; yet the two countries have an IWT intact. One of the reasons for this is that the IWT has not impeded their water requirements, at least till now; however, one cannot be certain about the IWT in future.

The second argument made in this book is that the roots of most of the transboundary water-related disputes in South Asia lie in the Partition of British India in 1947. Partition of territory by the Boundary Commission in 1947 also divided the rivers and canal system developed by the British imperialists during their rule. As mentioned in Chapters 2 and 3, it disrupted the age old interdependent hydraulic system. Both partition of the territories and the rivers are still being debated.

The erstwhile princely state of Jammu & Kashmir (J&K) is being claimed and counter-claimed by both India and Pakistan. In 1947 the king of J&K

had the option to become a part of either India or Pakistan. As he was taking his time to make decision, the kingdom was attacked by the Pakistan Army backed tribal intruders. They were first resisted by the locals in the Kashmir valley. As they made deep advances, the king, Hari Singh, requested the Indian government to help. India first took his signature on an accession paper and then sent its Army to throw away the intruders. The Indian Army pushed back the intruders and stopped after the issue was taken to the United Nations Security Council (UNSC) by India in 1948. After ceasefire, India has two-third parts of the J&K while some parts are with Pakistan and China. The UNSC asked for a plebiscite which has never happened, and it remains a matter of dispute between India and Pakistan. In 1972 under the Shimla agreement and in 1999 under the Lahore Declaration, India and Pakistan made it a bilateral issue. Since 1989 the Indian side of J&K has also witnessed a rise of militancy. In 2019, as India revoked special status to J&K under Article 370, there has been an uproar in Pakistan.

J&K is one of the water rich regions in South Asia. Besides ideological reasons, as mentioned in Chapter 2 of this book, Pakistan claims it because the western rivers of the IRS flow through the Kashmir valley into Pakistan. In its side of J&K and the Gilgit-Baltistan (GB) Pakistan has commissioned and is in the process of building HEPs. The electricity generated from those HEPs would likely meet some of the electricity demands of the Pakistani provinces. As these projects give little or almost no benefits to the local population, they have been opposed by the people from GB. Even India objects to such projects because it considers the Pakistan side of J&K and GB as its own area which have been under the illegal occupation of Pakistan since 1947–48.

On the Indian side of J&K, India has developed HEP projects on the shared waters. These projects have been contested by Pakistan, as some of them are on the western rivers whose majority of quantity of waters have been allocated to Pakistan under the IWT. Even people from Kashmir valley who feel the heat of the coercive power of the Indian State since Armed Forces Special Powers Act was clamped in J&K in 1990 believe that their water resources are, largely, being used by the Indian State to benefit others. For many Kashmiri, the author talked with, India's National Hydroelectric Power Corporation limited, exploits their water resources.

Like India and Pakistan, India and Bangladesh had also territorial disputes over enclaves and Adverse Possessions of lands. The dispute was there since 1947. In 1971 Bangladesh was liberated and in 1974 the two countries agreed on a Land Boundary Agreement (LBA). This was finally approved in 2011 after a protocol on it was signed between them. Later, in 2015 after the Indian parliament ratified it, LBA was implanted. In this exercise India gave more land to Bangladesh. Now the disputes have been resolved. Earlier in 2014, after the Permanent Court of Arbitration's judgement, the two countries resolved their maritime boundary issues.

In East Pakistan (now Bangladesh), boundary lines were even drawn in the middle of the border rivers. As discussed in Chapter 3, tribunal was set up to interpret and re-interpret the Radcliffe commission's decisions, however, disputes could not be resolved.

Most of the historical questions and tensions have not been resolved between the countries who once constituted British India. There is a long shadow of partition under which India, Pakistan, and Bangladesh are living. One such issue is of citizen and citizenship. To deport "outsiders" and non-citizens living "illegally" in Assam, in 2015, Indian government decided to prepare National Register of Citizens (NRC). In 2019, a final list of people non-eligible for Indian citizenship under NRC was published. Hyphenated with this is the Citizenship Amendment Act (CAA) of 2019 under which the government of India has decided to give citizenship to non-Muslim persecuted minorities from Pakistan, Afghanistan, and Bangladesh. This religion-based citizenship and objective of the NRC is mainly seen as targeting minorities, especially the Muslims.

The whole exercise under NRC and CAA may harm India's relationships with Bangladesh and Nepal. When NRC was carried out in Assam, Bangladesh was the centre of the entire debate. The objective of the NRC, as argued by the Union Government of India and the Assam government, was to detect and deport the illegally living Bangladeshi in Assam to Bangladesh. Conversely, Bangladesh's official stand is that none of its citizens are living illegally in India. Bangladesh states that all those who took shelter after the Pakistani Army began unleashing violence on the Bengali-speaking population from then East Pakistan had already returned after Bangladesh was liberated in 1971. On the NRC, during her visit to India in October 2019, the Prime Minister of Bangladesh, Sheikh Hasina, said: "I have spoken with PM Modi and I am satisfied".[8] Bangladesh's Foreign Secretary, Shahidul Haque, who, speaking on Hasina's visit to India, said: "Our (India-Bangladesh) relationship is currently the best of the best, and the relationship is extremely warm and friendly but at the same time, we are keeping our eyes quite open".[9] On being asked about the statements made by a few Indian leaders on deporting people who, finally, cannot be proven as Indian citizens, Haque said: "This (NRC) is internal matter (of India) and continue to believe in that and I think we shouldn't make a crisis out of nothing at this stage and we should be able to wait and see".[10]

Soon after CAA was passed, Abdul Momen cancelled his scheduled visit to India. This was followed by the cancellation of Bangladesh's Home Minister's Asaduzzaman Khan's visit and postponement of the two meetings on river management between the officials from the two countries in New Delhi.

Besides, according to media reports, between 60,000 and 100,000 Gurkhas living for centuries in Assam, have not found their names in the NRC list in Assam.[11] Most of these Gurkhas have their roots in Nepal but came and settled in India in different phases. In addition to Gurkhas, there are many other people of Nepali origin living in different parts of India from centuries. If NRC takes place for the entire country, many of them will likely to be excluded. This exclusion will have impact on India's bilateral relations with Nepal.

Hence, the NRC-CAA may have an indirect impact on water relations of India with Nepal and Bangladesh. This is because, as maintained in this book, countries having good bilateral relationships manage their trans-boundary related water issues in a better way.

The third argument in this book is about concerns over India aided HEPs in Nepal and Bhutan. Almost all such concerns have been examined in Chapters 4 and 5 of this book. To emerge from the shadow of India, Nepal has always tried to bring China so that the relationship with India could be balanced. In recent years, China-Nepal relations have improved a lot. Unlike Nepal, Bhutan remains dependent on India, though a section of Bhutanese calls on their government to explore relations with other countries, mainly China.

One of the reasons why transboundary water disputes occur and why cooperation is not reached or difficult to reach is the water-related stress riparian States face. In South Asia, almost all countries are facing water-related problems, but the degrees are different.

Highlighting the water situation in India, in June 2018, India's National Institution for Transforming India, came out with a report saying that India is facing its "worst water crisis in its history and millions of lives and livelihoods are under threat".[12] By 2030, the country's water demand is projected to be twice the available supply.[13]

India's water resources are unevenly distributed. To address India's water woes, the 1960s then-Union Minister of State for Power and Irrigation, K L Rao, spoke about the plan to link the Ganges River with the Cauvery River through a 2,640-kilometre long canal.[14] By the 1970s there was a plan to set up "national river grid" by which the surplus waters of the Ganga and Brahmaputra were to be diverted to the central and southern states. In 1974 an Air force pilot Dinshaw J Dastur submitted a proposal to the government of India. He suggested the construction of a 4,200-kilometres-long Himalayan canal and 9,300-kilometres-long southern canal to be linked up at New Delhi and Patna. Captain Dastur's proposal was popularly referred to as the Garland Canal.[15] The government of India prepared its own plan in 1980. The first step in this direction was setting up the National Water Development Agency (NWDA) in 1982 to carry out a detailed study of the river interlinking. The NWDA study supported interlinking of rivers to address the issue of floods in one region while drought in the other. In 2000 the Supreme Court of India was brought in to decide over the Interlinking of Rivers (ILR). The SC in its order directed the government of India to speed up the implementation of the ILR and complete it by 2016.[16] Consequently, the ILR began, though at a slow pace. The present government, under Indian Prime Minister Narendra Modi, has given it a push by providing much-required US $87 billion to the project.[17] This project aims to create 30 links across India, divided in two components – Himalayan (14 links) and Peninsular (16 links) – as well as some additional intra-state links.[18] Some of the rivers in this project are transboundary rivers which make the riparian States of India express a fear that once operationalised India may divert flows from these rivers to satisfy its water interests.

One of the proposed ILR project is to link Manas-Sankosh-Teesta-Ganga. It envisages a diversion of 43 BCM of surplus water of Manas, Sankosh, and intermediate rivers, for augmenting the flow of Ganga and provide 14 BCM of water in Mahanadi basin for further diversion to South through Peninsular link system.[19] Sankosh River, known as Punatsang Chu in Bhutan, is a transboundary river between India and Bhutan. Teesta is a transboundary river between India and Bangladesh. In the case of Sankosh, India is a lower riparian while it is upper riparian to Teesta. The concern is that once operationalised India may demand more waters from River Sankosh and may release less waters from Teesta to Bangladesh. As this is not a suitable place to discuss success or pitfalls of the ILR, one cannot put forward detailed arguments, however, it is a fact that such fears remain in India's two riparian States on this linking project. In such a situation, there is a need to take steps to address such fears.

On the status of the water availability in India and its South Asian riparian States, the World Bank's report titled *Climate Risks and Solutions: Adaptation Frameworks for Water Resources Planning, Development and Management in South Asia* 2017 has many stories to tell. One of the findings of the report is that the India's South Asian riparian Sates have to depend a lot on the outside river waters in the coming years. The reports define the Total Renewable Water Resource (TRWR) which is water that originates within each country and is a function of internal renewable water resources (IRWR), and water that flows from neighbouring countries as surface water entering the country (SWEC). Then it maintains that in South Asia the contribution of SWEC to TRWR varies from 0% (Bhutan) to > 90% (Bangladesh). The higher the SWEC the greater the dependence on inflows from other nations, making the receiving country more vulnerable to factors outside its control.[20] Higher dependence on the water from outside of nation would make the country vulnerable and also escalate the water disputes in the region. This change, the report maintains, is primarily because of factors like climate change, increase in demand of water, etc.[21]

The sad reality of India's South Asian transboundary rivers is that there are more disputes over them than cooperation. Despite having treaties and agreements, the countries, one or the other, violate or accused for violating of the provisions in those treaties and arrangements. In such scenario cooperation between India and its South Asian neighbours on the trans-boundary river waters could be possible:

(a) If countries de-link their waters from the day-to-day political affairs. For example, the discussions on IRS water sharing must be de-linked from both historic grievances and from the other Kashmir related issues.[22]

(b) There is a lack of imagination over a benefits-sharing paradigm due to cooperation over transboundary rivers water. In the transboundary water resources sense, benefit-sharing refers to a paradigm or policy tool that identifies the gains of interstate cooperation beyond merely the sharing of water but incorporates the sharing of opportunities that water brings to a country, a basin, and a region.[23] To cooperate such benefits-sharing paradigm has to be sketched.

(c) There is a need to change narratives of bilateral political relationships. A large section of the population in India looks at their small South Asian neighbours as "inconsequential" countries. Some even consider Nepal and Bhutan as an extended part of India while Pakistan is regarded as "enemy", and Bangladesh is always being accused of sending illegal people into the Indian territories. Likewise, narratives among a large section of population from the neighbouring countries too are anti-India in tone. Often, India is seen as a bullying power forcing them and "exploit" their resources. The change in narratives would certainly improve the bilateral relations further.

(d) Regional organisation could play a role but the South Asian Association for Regional Cooperation is in a defunct situation since 2016, after India decided to not participate in the Summit level meet which was to be held at Islamabad. The reason for it was in September that year, 20 Indian soldiers were killed in J&K in an attack by the militants. Other countries gave their own respective reasons to not participate in that year's SAARC summit.

(e) As most of the major river systems of South Asia originate in China, it is going to be an important factor in the future of South Asia's water security. China itself faces water stress and to meet its demand it is being accused by India of diverting water flows of the Rivers Brahmaputra and Sutlej (part of the IRS). Any water-related activities in the upper riparian, China, also affects India's release of this river's waters to Bangladesh. In such situation, to deal with their demand-related issue, the South Asian riparian States better join together and discuss the issue with China. As pointed in Chapter 1, China's riparian position to many important South Asian Rivers and its deep engagements in water sectors of Bangladesh, Pakistan and Nepal make it a party which can influence the water contest between India and its riparian neighbours.[24]

(f) Water is also, looked as a source of power which one riparian State can enjoy over the others. On transboundary waters, the relatively powerful riparian State tries to dictate its terms over water rich smaller States. In South Asia, as discussed in Chapters 3, 4, and 5, India is being accused by many people in Nepal, Bangladesh, and Bhutan for treaty and agreements favouring India at the cost of their water resources. In cases where the riparian countries are equally powerful in terms of military and economy, disputes tend to escalate, as discussed in Chapter 2. Pakistan is not more powerful than India in terms of conventional military force and in economic terms, but possession of nuclear weapons gives it some advantage to contest India's position on the shared rivers. For any probable cooperation this power element has to be replaced with equitable and reasonable distribution of waters from the transboundary rivers. This replacement has to reflect on water sharing documents, and more on ground.

As water scarcity looms large on India and its co-riparian South Asian States, all of them have to discuss, plan, and decide on the ways to meet such foreseeable challenges. Unless they cooperate, South Asia's transboundary river waters will remain contested.

Notes

1 "Smart Fence Along Bangla Border in Assam by July 2020: BSF DG" *The Economic Times* 2019, 29 December.https://m.economictimes.com/news/defence/smart-fence-along-riverine-bangla-border-in-assam-by-july-2020-bsf-dg/articleshow/73020134.cms. Accessed on 18 January 2020.
2 Singh, Shiv Sahay and Subhojit Bagchi. (2018, 26 May). "Want to Settle All Issues in a Friendly Ambience: Hasina". *The Hindu*. www.thehindu.com/news/national/want-to-settle-all-issues-in-a-friendly-ambience-hasina/article23994739.ece?homepage=true. Accessed on 26 May 2018.
3 "Mamata Hasina Meet a Courtesy Call". *United News of India*, 2018, 26 May. www.uniindia.com/mamata-hasina-meet-a-courtesy-call/states/news/1242802.html. Accessed on 27 May 2018.
4 "Let's Find a Mutually Acceptable Formula". *The Daily Star*, 2019, 21 August. www.thedailystar.net/frontpage/news/lets-find-mutually-acceptable-formula-1788139. Accessed on 22 August 2019.
5 Ibid.
6 Ibid.
7 Ibid.
8 Bagchi, Indrani. (2019, 4 October). "Modi Assured NRC Process Won's Affect Bangladesh". *The Times of India*. https://timesofindia.indiatimes.com/india/modi-assured-nrc-process-wont-affect-bangladesh-hasina/articleshow/71432172.cms. Accessed on 5 October 2019.
9 Roche, Elizabeth., (2019, 6 October). "Bangladesh Reassured on NRC, Shouldn't Make a Crisis OUT of Nothing". *The Mint*.
10 Ibid.
11 Naqvi, Sadiq. (2019, 24 September). Gorkhas Up in Arms Against Omissions From Assam NRC List". *Hindustan Times*. https://www.hindustantimes.com/india-news/gorkhas-up-in-arms-against-omissions-from-assam-nrc-list/story-FHSiSH0ZSC4g3FXflwfinN.html. Accessed on 28 November 2019.

12 NITI Aayog, Government of India. "Composite Water Management Index: A Tool for Water Management". http://niti.gov.in/writereaddata/files/document_publication/2018-05-18-Water-Index-Report_vS8-compressed.pdf. Accessed on 25 June 2018, p. 15.
13 Ibid.
14 D'Souza, Rohan. (2003, 6 September). "Supply-Side Hydrology in India". *Economic &Political Weekly*, pp. 3785–3790.
15 Ibid.
16 Narain, Sunita. "Grand Distraction Called Interlinking". Centre for Science and Environment. www.cseindia.org/content/grand-distraction-called-river-inter linking. Accessed on 11 December 2017.
17 "PM Modi's $87 Billion River-Linking Gamble Set to Take Off as Floods Hit India". *India Today*. 2017, 1 September. https://www.indiatoday.in/india/mad hya-pradesh/story/narendra-modi-river-linking-plan-floods-hit-india-ganges-go davari-mahanadi-1035664-2017-09-01. Accessed on 12 October 2017.
18 Ministry of jal Shakti, Water Resources & Ganga Rejuvenation, Government of India. "Inter Linking of Rivers". http://mowr.gov.in/schemes-projects-pro grammes/schemes/interlinking-rivers. Accessed on 16 September 2019.
19 Ibid.
20 Climate Risks and Solutions: Adaptation Frameworks for Water Resources Planning, Development and Management in South Asia. (2017). www.iwmi.cgiar. org/wp-content/uploads/2017/06/Climate-Risks-and-Solutions-Adaptation-Frameworks-for-Water-Resources-Planning-Development-and-Management-in-South-Asia.pdf. Accessed on 18 July 2018, p. 9.
21 See Ibid.
22 Briscoe, John. (2010, 11–17 December). "Troubled Waters: Can a Bridge Be Built Over River Indus". *Economic &Political Weekly*, Vol. xlv, No. 50, pp. 28–32.
23 Jacobs, Inga M. (2012). *The Politics of Water in Africa: Norms, Environmental Regions and Transboundary Cooperation in the Orange-Senqu and Nile Rivers*. London, and New York: Continuum.
24 Ranjan, Amit. (2019, December). "China's Infrastructure Projects in South Asia: An Appraisal". *Contemporary Chinese Political Economy and Strategic Relations; Kaohsiung*, Vol. 5, No. 3, pp. 1079–1110.

Appendices

Appendix 1
Indus Waters Treaty

Karachi

PREAMBLE

The Government of India and the Government of Pakistan, being equally desirous of attaining the most complete and satisfactory utilisation of the waters of the Indus system of rivers and recognising the need, therefore, of fixing and delimiting, in a spirit of goodwill and friendship, the rights and obligations of each in relation to the other concerning the use of these waters and of making provision for the settlement, in a cooperative spirit, of all such questions as may hereafter arise in regard to the interpretation or application of the provisions agreed upon herein, have resolved to conclude a Treaty in furtherance of these objectives, and for this purpose have named as their plenipotentiaries:

THE GOVERNMENT OF INDIA:
Shri JAWAHARLAL NEHRU,
Prime Minister of India,
and
THE GOVERNMENT OF PAKISTAN
Field Marshal MOHAMMAD AYUB KHAN, HP., H.J.,
President of Pakistan;

who, having communicated to each other their respective Full Powers and having found them in good and due form, have agreed upon the following Articles and Annexures:-

Article I

Definitions
As used in this Treaty:

1 The terms "Article" and "Annexure" mean respectively an Article of, and an Annexure to, this Treaty.

Except as otherwise indicated, references to Paragraphs are to the paragraphs in the Article or in the Annexure in which the reference is made.

2 The term "Tributary" of a river means any surface channel whether in continuous or intermittent flow and by whatever name called, whose waters in the natural course would fall into that river, e.g. a tributary, a torrent, a natural drainage, an artificial drainage, a nadi, a nallah, a nai, a khad, a cho. The term also includes any sub-tributary or branch or subsidiary channel, by whatever name called, whose waters, in the natural course, would directly or otherwise flow into that surface channel.

3 The term "The Indus", "The Jhelum", "The Chenab", "The Ravi", "The Beas" or "The Sutlej" means the named river (including Connecting Lakes, if any) and all its Tributaries: Provided however that

I   none of the rivers named above shall be deemed to be a Tributary;
II   The Chenab shall be deemed to include the river Panjnad; and
III the river Chandra and the river Bhaga shall be deemed to be Tributaries of The Chenab.

4 The term "Main" added after Indus, Jhelum, Chenab, Sutlej, Beas or Ravi means the main stem of the named river excluding its Tributaries, but including all channels and creeks of the main stem of that river and such Connecting Lakes as form part of the main stem itself. The Jhelum Main shall be deemed to extend up to Verinag, and the Chenab Main up to the confluence of the river Chandra and the river Bhaga.

5 The term "Eastern Rivers" means The Sutlej, The Beas and The Ravi taken together.

6 The term "Western Rivers" means The Indus, The Jhelum and The Chenab taken together.

7 The term "the Rivers" means all the rivers, The Sutlej, The Beas, The Ravi, The Indus, The Jelum and The Chenab.

8 The term "Connecting Lake" means any lake which receives water from, or yields water to, any of the Rivers; but any lake which occasionally and irregularly receives only the spill of any of the Rivers and returns only the whole or part of that spill is not a Connecting Lake.

9 The term "Agricultural Use" means the use of water for irrigation, except for irrigation of household gardens and public recreational gardens.

10 The terms "Domestic Use" means the use of water for

(a) drinking, washing, bathing, recreation, sanitation (including the conveyance and dilution of sewage and of industrial and other wastes), stock and poultry, and other like purposes;
(b) household and municipal purposes (including use for household gardens and public recreational gardens); and
(c) industrial purposes (including mining, milling and other like purposes); but the term does not include Agricultural Use or use for the generation of hydro-electric power.

11 The term "Non-Consumptive Use" means any control or use of water for navigation, floating of timber or other property, flood protection or flood control, fishing or fish culture, wild life or other like beneficial purposes, provided that, exclusive of seepage and evaporation of water incidental to the control or use, the water (undiminished in volume within the practical range of measurement) remains in, or is returned to, the same river or its Tributaries; but the term does not include Agricultural Use or use for the generation of hydro-electric power.

12 The term "Transition Period" means the period beginning and ending as provided in Article 11(6).

13 The term "Bank" means the International Bank for Reconstruction and Development.

14 The term "Commissioner" means either of the Commissioners appointed under the provisions of Article VIII(1) and the term "Commission" means the Permanent Indus Commission constituted in accordance with Article VIII(3).

15 The term "interference with the waters" means:

   (a) Any act of withdrawal therefrom; or
   (b) Any man-made obstruction to their flow which causes a change in the volume (within the practical range of measurement) of the daily flow of the water: Provided however that an obstruction which involves only an insignificant and incidental change in the volume of the daily now, for example, fluctuations due to afflux caused by bridge piers or a temporary by-pass, etc., shall not be deemed to be an interference with the waters.

16 The term "Effective Date" means the date on which this Treaty takes effect in accordance with the provisions of Article XII, that is, the first of April 1960.

Article II

Provisions Regarding Eastern Rivers

1 All the waters of the Eastern Rivers shall be available for the unrestricted use of India, except as otherwise expressly provided in this Article.

2 Except for Domestic Use and Non-Consumptive Use, Pakistan shall be under an obligation to let flow, and shall not permit any interference with, the waters of the Sutlej Main and the Ravi Main in the reaches where these rivers flow in Pakistan and have not yet finally crossed into Pakistan. The Points of final crossing are the following: (a) near the new Hasta Bund upstream of Suleimanke in the case of the Sutlej Main, and (b) about one and a half miles upstream of the syphon for the B-R-B-D Link in the case of the Ravi Main.

3   Except for Domestic Use, Non-Consumptive Use and Agricultural Use (as specified in Annexure B), Pakistan shall be under an obligation to let flow, and shall not permit any interference with, the waters (while flowing in Pakistan) of any Tributary which in its natural course joins the Sutlej Main or the Ravi Main before these rivers have finally crossed into Pakistan.

4   All the waters, while flowing in Pakistan, of any Tributary which, in its natural course, joins the Sutlej Main or the Ravi Main after these rivers have finally crossed into Pakistan shall be available for the unrestricted use of Pakistan: Provided however that this provision shall not be construed as giving Pakistan any claim or right to any releases by India in any such Tributary. If Pakistan should deliver any of the waters of any such Tributary, which on the Effective Date joins the Ravi Main after this river has finally crossed into Pakistan, into a reach of the Ravi Main upstream of this crossing, India shall not make use of these waters; each Party agrees to establish such discharge observation stations and make such observations as may be necessary for the determination of the component of water available for the use of Pakistan on account of the aforesaid deliveries by Pakistan, and Pakistan agrees to meet the cost of establishing the aforesaid discharge observation stations and making the aforesaid observations.

5   There shall be a Transition Period during which, to the extent specified in Annexure H, India shall

I   limit its withdrawals for Agricultural Use,
II   limit abstractions for storages, and
III make deliveries to Pakistan from the Eastern Rivers.

6   The Transition Period shall begin on 1st April 1960 and it shall end on 31st March 1970, or, if extended under the provisions of Part 8 of Annexure H, on the date up to which it has been extended. In any event, whether or not the replacement referred to in Article IV(1) has been accomplished, the Transition Period shall end not later than 31st March 1973.

7   If the Transition Period is extended beyond 31st March 1970, the Provisions of Article V(5) shall apply.

8   If the Transition Period is extended beyond 31st March 1970, the provisions of Paragraph (5) shall apply during the period of extension beyond 31st March 1970.

9   During the Transition Period, Pakistan shall receive for unrestricted use the waters of the Eastern Rivers which are to be released by India in accordance with the provisions of Annexure H. After the end of the Transition Period, Pakistan shall have no claim or right to releases by India of any of the waters of the Eastern Rivers. In case there are any releases, Pakistan shall enjoy the unrestricted use of the waters so released after they have finally crossed into Pakistan: Provided that in

the event that Pakistan makes any use of these waters, Pakistan shall not acquire any right whatsoever, by prescription or otherwise, to a continuance of such releases or such use.

Article III

Provisions Regarding Western Rivers

1 Pakistan shall receive for unrestricted use all those waters of the Western Rivers which India is under obligation to let flow under the provisions of Paragraph (2).

2 India shall be under an obligation to let flow all the waters of the Western Rivers, and shall not permit any interference with these waters, except for the following uses, restricted (except as provided in item (c) (11) of Paragraph 5 of Annexure (C) in the case of each of the rivers, The Indus, The Jhelum and The Chenab, to the drainage basin thereof

   (a) Domestic Use;
   (b) Non-Consumptive Use;
   (c) Agricultural Use, as set out in Annexure C; and
   (d) Generation of hydro-electric power, as set out in Annexure D.

3 Pakistan shall have the unrestricted use of all waters originating from sources other than the Eastern Rivers which are delivered by Pakistan into The Ravi or The Sutlej, and India shall not make use of these waters. Each Party agrees to establish such discharge observation stations and make such observations as may be considered necessary by the Commission for the determination of the component of water available for the use of Pakistan on account of the aforesaid deliveries by Pakistan.

4 Except as provided in Annexure D and E, India shall not store any water of, or construct any storage works on, the Western Rivers.

Article IV

Provisions Regarding Eastern Rivers and Western Rivers

1 Pakistan shall use its best endeavours to construct and bring into operation, with due regard to expedition and economy, that part of a system of works which will accomplish the replacement, from the Western Rivers and other sources, of water supplies for irrigation canals in Pakistan which, on 15th August 1947, were dependent on water supplies from the Eastern Rivers.

2 Each Party agrees that any Non-Consumptive Use made by it shall be so made as not to materially change, on account of such use, the flow in any channel to the prejudice of the uses on that channel by the other Party under the provisions of this Treaty. In executing any scheme of

flood protection or flood control each Party will avoid, as far as practicable, any material damage to the other Party, and any such scheme
carried out by India on the Wejern Rivers shall not involve any use of
water or any storage in addition to that provided under Article III.

3 Nothing in this Treaty shall be construed as having the effect of prevent-
ing either Party from undertaking schemes of drainage, river training,
conservation of soil against erosion and dredging, or from removal of
stones, gravel or sand from the beds of the Rivers: Provided that

    (a) in executing any of the schemes mentioned above, each Party will
avoid, as far as practicable, any material damage to the other Party;

    (b) any such scheme carried out by India on the Western Rivers shall
not involve any use of water or any storage in addition to that provided under Article III;

    (c) except as provided in Paragraph (5) and Article VII(l)(b), India shall
not take any action to increase the catchment area, beyond the area
on the Effective Date, of any natural or artificial drainage or drain
which crosses into Pakistan, and shall not undertake such construction or remodelling of any drainage or drain which so crosses or
falls into a drainage or drain which so crosses as might cause material damage in Pakistan or entail the construction of a new drain or
enlargement of an existing drainage or drain in Pakistan; and

    (d) should Pakistan desire to increase the catchment area, beyond the
area on the Effective Date, of any natural or artificial drainage or
drain, which receives drainage waters from India, or, except in an
emergency, to pour any waters into it in excess of the quantities
received by it as on the Effective Date, Pakistan shall, before undertaking any work for these purposes, increase the capacity of that
drainage or drain to the extent necessary so as not to impair its
efficacy for dealing with drainage waters received from India as on
the Effective Date.

4 Pakistan shall maintain in good order its portions of the drainages mentioned below with capacities not less than the capacities as on the Effective Date

    Hudiara Drain
  I   Kasur Nala
  II   Salimshah Drain
  III Fazilka Drain.

5 If India finds it necessary that any of the drainages mentioned in Paragraph (4) should be deepened or widened in Pakistan, Pakistan agrees
to undertake to do so as a work of public interest, provided India agrees
to pay the cost of the deepening or widening.

6 Each Party will use its best endeavours to maintain the natural channels
of the Rivers, as on the Effective Date, in such condition as will avoid,

as far as practicable, any obstruction to the flow in these channels likely to cause material damage to the other Party.

7 Neither Party will take any action which would have the effect of diverting the Ravi Main between Madhopur and Lahore, or the Sutlej Main between Harike and Suleimanke, from its natural channel between high banks.

8 The use of the natural channels of the Rivers for the discharge of flood or other excess waters shall be free and not subject to limitation by either Party, and neither Party shall have any claim against the other in respect of any damage caused by such use. Each Party agrees to communicate to the other Party, as far in advance as practicable, any information it may have in regard to such extraordinary discharges of water from reservoirs and flood flows as may affect the other Party.

9 Each Party declares its intention to operate its storage dams, barrages and irrigation canals in such manner, consistent with the normal operations of its hydraulic systems, as to avoid, as far as feasible, material damage to the other Party.

10 Each Party declares its intention to prevent, as far as practicable, undue pollution of the waters of the Rivers which might affect adversely uses similar in nature to those to which the waters were put on the Effective Date, and agrees to take all reasonable measures to ensure that, before any sewage or industrial waste is allowed to flow into the Rivers, it will be treated, where necessary, in such manner as not materially to affect those uses:

Provided that the criterion of reasonableness shall be the customary practice in similar situations on the Rivers.

11 The Parties agree to adopt, as far as feasible, appropriate measures for the recovery, and restoration to owners, of timber and other property floated or floating down the Rivers, subject to appropriate charges being paid by the owners.

12 The use of water for industrial purposes under Articles 11(2), 11(3) and III (2) shall not exceed :

a in the case of an industrial process known on the Effective Date, such quantum of use as was customary in that process on the Effective Date;

b in the case of an industrial process not known on the Effective Date:

(i) such quantum of use as was customary on the Effective Date in similar or in any way comparable industrial processes; or

(ii) if there was no industrial process on the Effective Date similar or in any way comparable to the new process, such quantum of use as would not have a substantially adverse effect on the other Party.

13   Such part of any water withdrawn for Domestic Use under the provisions of Articles 11(3) and 111(2) as is subsequently applied to Agricultural Use shall be accounted for as part of the Agricultural Use specified in Annexure B and Annexure C respectively; each Party will use its best endeavours to return to the same river (directly or through one of its Tributaries) all water withdrawn therefrom for industrial purposes and not consumed either in the industrial processes for which it was withdrawn or in some other Domestic Use.

14   In the event that either Party should develop a use of the waters of the Rivers which is not in accordance with the provisions of this Treaty, that Party shall not acquire by reason of such use any right, by prescription or otherwise, to a continuance of such use.

15   Except as otherwise required by the express provisions of this Treaty, nothing in this Treaty shall be construed as affecting existing territorial rights over the waters of any of the Rivers or the beds or banks thereof, or as affecting existing property rights under municipal law over such waters of beds or banks.

Article V

Financial Provisions

1   In consideration of the fact that the purpose of part of the system of works referred to in Article IV(1) is the replacement, from the Western Rivers and other sources, of water supplies for irrigation canals in Pakistan which, on 15th August 1947, were dependent on water supplies from the Eastern Rivers, India agrees to make a fixed contribution of Pounds Sterling 62,060,000 towards the costs of these works. The amount in Pounds Sterling of this contribution shall remain unchanged irrespective of any alteration in the par value of any currency.

2   The sum of Pounds Sterling 62,060,000 specified in Paragraph (1) shall be paid in ten equal annual instalments on the Ist of November of each year. The first of such annual instalments shall be paid on lst November 1960, or if the Treaty has not entered into force by that date, then within one month after the Treaty enters into force.

3   Each of the instalments specified in Paragraph (2) shall be paid to the Bank for the credit of the Indus Basin Development Fund to be established and administered by the Bank, and payment shall be made in Pounds Sterling, or in such other currency or currencies as may from time to time be agreed between India and the Bank.

4   The payments provided for under the provisions of Paragraph (3) shall be made without deduction or set-off on account of any financial claims of India on Pakistan arising otherwise than under the provisions of this Treaty: Provided that this provision shall in no way absolve Pakistan

from the necessity of paying in other ways debts to India which may be outstanding against Pakistan.

5   If, at the request of Pakistan, the Transition Period is extended in accordance with the provisions of Article 11(6) and of Part 8 of Annexure H, the Bank shall thereupon pay to India out of the Indus Basin Development Fund the appropriate amount specified in the Table below:

Period of Aggregate	*Payment to India*
Extension Payment of Transition Period	
One year	stg 3,125,000
Two years	stg 6,406,250
Three years	stg 9,850,000

6   The provisions of Article IV(1) and Article V(1) shall not be construed as conferring upon India any right to participate in the decisions as to the system of works which Pakistan constructs pursuant to Article IV(1) or as constituting an assumption of any responsibility by India or as an agreement by india in regard to such works.

7   Except for such payments as are specifically provided for in this Treaty, neither Party shall be entitled to claim any payment for observance of the provisions of this Treaty or' to make any charge for water received from it by the other Party.

Article VI

Exchange of Data

1   The following data with respect to the flow in, and utilisation of the waters of, the Rivers shall be exchanged regularly between the Parties:

a   Daily (or as observed or estimated less frequently) gauge and discharge data relating to flow of the Rivers at all observation sites.
b   Daily extractions for or releases from reservoirs.
c   Daily withdrawals at the heads of all canals operated by government or by a government agency (hereinafter in this Article called canals), including link canals.
d   Daily escapages from all canals, including link canals.
e   Daily deliveries from link canals.

These data shall be transmitted monthly by each Party to the other as soon as the data for a calendar month have been collected and tabulated, but not later than three months after the end of the month to which they relate:

Provided that such of the data specified above as are considered by either Party to be necessary for operational purposes shall be supplied daily or at less frequent intervals, as may be requested. Should one Party request the supply of any of these-data by telegram, telephone, or wireless, it shall reimburse the other Party for the cost of transmission.

2   If, in addition to the data specified in Paragraph (1) of this Article, either Party requests the supply of any data relating to the hydrology of the Rivers, or to canal or reservoir operation connected with the Rivers, or to anv provision of this Treaty, such data shall be supplied by the other Party to the extent that these are available.

Article VII

Future Co-operation

1   The two Parties recognize that they have a common interest in the optimum development of the Rivers, and, to that end, they declare their intention to co-operate, by mutual agreement, to the fullest possible extent. In particular:

   a   Each Party, to the extent it considers practicable and on agreement by the other Party to pay the costs to be incurred, will, at the request of the other Party, set up or install such hydrologic observation stations within the drainage basins of the Rivers, and set up or install such meteorological observation stations relating thereto and carry out such observations thereat, as may be requested, and will supply the data so obtained.
   b   Each Party, to the extent it considers practicable and on agreement by the other Party to pay the costs to be incurred, will, at the request of the other Party, carry out such new drainage works as may be required in connection with new drainage works of the other Party.
   c   At the request of either Party, the two Parties may, by mutual agreement, co-operate in undertaking engineering works on the Rivers.

The formal arrangements, in each case, shall be as agreed upon between the Parties.

2   If either Party plans to construct any engineering work which would cause interference with the waters of any of the Rivers and which, in its opinion, would affect the other Party materially, it shall notify the other Party of its plans and shall supply such data relating to the work as may be available and as would enable the other Party to inform itself of the nature, magnitude and effect of the work. If a work would cause interference with the waters of any of the Rivers but would not, in the

opinion of the Party planning it, affect the other Party materially, nevertheless the Party planning the work shall, on request, supply the other Party with such data regarding the nature, magnitude and effect, if any, of the work as may be available.

Article VIII

Permanent Indus Commission

1 India and Pakistan shall each create a permanent post of Commissioner for Indus Waters, and shall appoint to this post, as often as a vacancy occurs, a person who should ordinarily be a high-ranking engineer competent in the field of hydrology and water-use. Unless either Government should decide to take up any particular question directly with the other Government, each Commissioner will be the representative of his Government for all. matters arising out of this Treaty, and will serve as the regular channel of communication on all matters relating to the implementation of the Treaty, and, in particular, with respect to

   a the furnishing or exchange of information or data provided for in the Treaty; and
   b the giving of any notice or response to any notice provided for in the Treaty.

2 The status of each Commissioner and his duties and responsibilities towards his Government will be determined by that Government.
3 The two Commissioners shall together form the Permanent Indus Commission.
4 The purpose and functions of the Commission shall be to establish and maintain co-operative arrangements for the, implementation of this Treaty, to promote co-operation between the Parties in the development of the waters of the Rivers and, in particular,

   a to study and report to the two Governments on any problem relating to the development of the waters of the Rivers which may be jointly referred to the Commission by the two Governments: in the event that a reference is made by one Government alone, the Commissioner of the other Government shall obtain the authorization of his Government before he proceeds to act on the reference;
   b to make every effort to settle promptly, in accordance with the provisions of Article IX(1), any question arising there under;
   c to undertake, once in every five years, a general tour of inspection of the Rivers for ascertaining the facts connected with various developments and works on the Rivers;
   d to undertake promptly, at the request of either Commissioner, a tour of inspection of such works or sites on the Rivers as may be

considered necessary by him for ascertaining the facts connected with those works or sites; and

e   to take, during the Transition Period, such steps as may be necessary for the implementation of the provisions of Annexure H.

5   The Commission shall meet regularly at least once a year, alternately in India and Pakistan. This regular annual meeting shall be held in November or in such other month as may be agreed upon between the Commissioners. The Commission shall also meet when requested by either Commissioner.

6   To enable the Commissioners to perform their functions in the Commission, each Government agrees to accord to the Commissioner of the other Government the same privileges and immunities as are accorded to representatives of member States to the principal and subsidiary organs of the United Nations under Sections 11, 12 and 13 of Article IV of the Convention on the Privileges and Immunities of the United Nations (dated 13th February, 1946) during the periods specified in those Sections. It is understood and agreed that these privileges and immunities are accorded to the Commissioners not for the personal benefit of the individuals themselves but in order to safeguard the independent exercise of their functions in connection with the Commission; consequently, the Government appointing the Commissioner not only has the right but is under a duty to waive the immunity of its Commissioner in any case where, in the opinion of the appointing Government, the immunity would impede the course of justice and can be waived without prejudice to the purpose for which the immunity is accorded.

7   For the purposes of the inspections specified in Paragraph (4) (c) and (d), each Commissioner may be accompanied by two advisers or assistants to whom appropriate facilities will be accorded.

8   The Commission shall submit to the Government of India and to the Government of Pakistan, before the first of June of every year, a report on its work for the year ended on the preceding 31st of March, and may submit to the two Governments other reports at such times as it may think desirable.

9   Each Government shall bear the expenses of its Commissioner and his ordinary staff. The cost of any special staff required in connection with the work mentioned in Article VII(1) shall be borne as provided therein.

10   The Commission shall determine its own procedures.

Article IX

Settlement of Differences and Disputes

1   Any question which arises between the Parties concerning the interpretation or application of this Treaty or the existence of any fact which, if established, might constitute a breach of this Treaty shall first be

examined by the Commission, which will endeavour to resolve the question by agreement.

2 If the Commission does not reach agreement on any of the questions mentioned in Paragraph (1), then a difference will be deemed to have arisen, which shall be dealt with as follows:

    a Any difference which, in the opinion of either Commissioner, falls within the provisions of Part I of Annexure F shall, at the request of either Commissioner, be dealt with by a Neutral Expert in accordance with the provisions of Part 2 of Annexure F;

    b If the difference does not come within the provisions of Paragraph (2) (a), or if a Neutral Expert, in accordance with the provisions of Paragraph 7 of Annexure F, has informed the Commission that, in his opinion, the difference, or a part thereof, should be treated as a dispute, then a dispute will be deemed to have arisen which shall be settled in accordance with the provisions of Paragraphs (3), (4) and (5):

Provided that, at the discretion of the Commission, any difference may either be dealt with by a Neutral Expert in accordance with the provisions of Part 2 of Annexure F or be deemed to be a dispute to be settled in accordance with the provisions of Paragraphs (3), (4) and (5), or may be settled in any other way agreed upon by the Commission.

3 As soon as a dispute to be settled in accordance with this and the succeeding paragraphs of this Article has arisen, the Commission shall, at the request of either Commissioner, report the fact to the two Governments, as early as practicable, stating in its report the points on which the Commission is in agreement and the issues in dispute, the views of each Commissioner on these issues and his reasons therefore.

4 Either Government may, following receipt of the report referred to in Paragraph (3), or if it comes to the conclusion that the report is being unduly delayed in the Commission, invite the other Government to resolve the dispute by agreement. In doing so it shall state the names of its negotiators and their readiness to meet with the negotiators to be appointed by the other Government at a time and place to be indicated by the other Government. To assist in these negotiations, the two Governments may agree to enlist the services of one or more mediators acceptable to them.

5 A Court of Arbitration shall be established to resolve the dispute in the manner provided by Annexure G

a upon agreement between the Parties to do so; or

    b at the request of either Party, if, after negotiations have begun pursuant to Paragraph (4), in its opinion the dispute is not, likely to be resolved by negotiation or mediation; or

    c at the request of either Party, if, after the expiry of one month following receipt by the other Government of the invitation referred to in Paragraph (4), that Party comes to the conclusion that the other Government is unduly delaying the negotiations.

5   The provisions of Paragraphs (3), (4) and (5) shall not apply to any difference while it is being dealt with by a Neutral Expert.

Article X

Emergency Provision

If, at any time prior to 31st March 1965, Pakistan should represent to the Bank that, because of, the outbreak of large-scale international hostilities arising out of causes beyond the control of Pakistan, it is unable to obtain from abroad the materials and equipment necessary for the completion, by 31st March 1973, of that part of the system of works referred to in Article IVU) which relates to the replacement referred to therein, (hereinafter referred to as the "replacement element") and if, after consideration of this representation in consultation with India, the Bank is of the opinion that

a   these hostilities are on a scale of which the consequence is that Pakistan is unable to obtain in time such materials and equipment as must be procured from abroad for the completion, by 31st March 1973, of the replacement element, and

b   since the Effective Date, Pakistan has taken all reasonable steps to obtain the said materials and equipment and, with such resources of materials and equipment as have been available to Pakistan both from within Pakistan and from abroad, has carried forward the construction of the replacement element with due diligence and all reasonable expedition,

the Bank shall immediately notify each of the Parties accordingly. The Parties undertake, without prejudice to the provisions of Article XII (3) and (4), that, on being so notified, they will forthwith consult together and enlist the good offices of the'Bank in their consultation, with a view to reaching mutual agreement as to whether or not, in the light of all the circumstances then prevailing, any modifications of the provisions of this Treaty are appropriate and advisable and, if so, the nature and the extent of the modifications.

Article XI

General Provisions

1   It is expressly understood that

a   this Treaty governs the rights and obligations of each Party in relation to the other with respect only to the use of the waters of the Rivers and matters incidental thereto; and b   nothing contained in this Treaty, and nothing arising out of the execution thereof, shall be construed as constituting a recognition or waiver (whether tacit, by implication or otherwise) of any rights or claims whatsoever of either of the Parties other than those rights or claims which are expressly recognized or waived in this Treaty.

Each of the Parties agrees that it will not invoke this Treaty, anything contained therein, or anything arising out of the execution thereof, in support of any of its own rights or claims whatsoever or in disputing any of the rights or claims whatsoever of the other Party, other than those rights or claims which are expressly recognized or waived in this Treaty.

2   Nothing in this Treaty shall be construed by the Parties as in any way establishing any general principle of law or any precedent.

3   The rights and obligations of each Party under this Treaty shall remain unaffected by any provisions contained in, or by anything arising out of the execution of, any agreement establishing the Indus Basin Development Fund.

Article XII

Final Provisions

1   This Treaty consists of the Preamble, the Articles hereof and Annexures A to H hereto, and may be cited as "The Indus Waters Treaty 1960".

2   This Treaty shall be ratified and the ratifications thereof shall be exchanged in New Delhi. It shall enter into force upon the exchange, of ratifications, and will then take effect retrospectively from the first of April 1960.

3   The provisions of this Treaty may from time to time be modified by a duly ratified treaty concluded for that purpose between the two Governments.

4   The provisions of this Treaty, or, the provisions of this Treaty as modified under the provisions of Paragraph (3), shall continue in force until terminated by a duly ratified treaty concluded for that purpose between the two Governments.

IN WITNESS WHEREOF the respective Plenipotentiaries have signed this Treaty and have hereunto affixed their seals.

DONE in triplicate in English at Karachi on this Nineteenth day of September 1960.

For the Government of India   For the Government of Pakistan
(Sd) JAWAHARLAL NEHRU   (Sd) MOHAMMAD AYUB KHAN
                                                        Field Marshal, H.P., H.J.

For the International Bank for Reconstruction and Development for the purposes specified in Articles V and X and Annexures F, G and H:
(Sd) W.A.B. ILIFF

Source: Ministry of Jal Shakti, Department of Water Resources, River Development & Ganga Rejuvenation (1960), Government of India, "The Indus Waters Treaty 1960". http://jalshakti-dowr.gov.in/sites/default/files/INDUS%20WATERS%20TREATY.pdf. Accessed on 18 July 2019.

Appendix 2

Treaty between the Government of the Republic of India and the Government of the People's Republic of Bangladesh on sharing of the Ganga/Ganges waters at Farakka

THE GOVERNMENT OF THE REPUBLIC OF INDIA AND THE GOVERNMENT OF THE PEOPLE'S REPUBLIC OF BANGLADESH.

DETERMINED to promote and strengthen their relations of friendship and good neighbourliness. INSPIRED by the common desire of promoting the well-being of their peoples,

BEING desirous of sharing by mutual agreement the waters of the international rivers flowing through the territories of the two countries and of making the optimum utilization of the water resources of their region in the fields of flood management, irrigation, river-basin development and generation of hydropower for the mutual benefit of the peoples of the two countries.

RECOGNISING that the need for making an arrangement for sharing of the Ganga/Ganges waters at Farakka in a spirit of mutual accommodation and the need for a solution to the long-term problem of augmenting the flows of Ganga/Ganges are in the mutual interests of the peoples of the two countries.

BEING desirous of finding a fair and just solution without affecting the rights and entitlements of either country other than those covered by this Treaty, or establishing any general principles of law or precedent.

HAVE AGREED AS FOLLOWS:-

Article I

The quantum of waters agreed to be released by India to Bangladesh will be at Farakka.

Article II

(i) The sharing between India and Bangladesh of the Ganga/Ganges waters at Farakka by ten days periods from the 1st of January to the 31st May every year will be with reference to the formula at Annexure I and an indicative schedule giving the implications of the sharing arrangement under Annexure I is at Annexure II.

(ii)   The indicative schedule at Annexure II; as referred to in sub para [1] above is based on 40 years (1949–1988) 10-day period average availability of water at Farakka. Every effort would be made by the upper riparian to protect flows of water at Farakka as in the 40 years average availability as mentioned above.

(iii) In the event flow at Farakka falls below 50,000 cusecs in any 10 – day period, the two governments will enter into immediate consultations to make adjustments on an emergency basis, in accordance with the principles of equity, fair play and no harm to either party.

Article III

The waters released to Bangladesh at Farakka under Article I shall not be reduced below Farakka except for reasonable uses of water, not exceeding 200 cusecs, by India between Farakka and the point on the Ganga/Ganges where both its banks are in Bangladesh.

Article IV

A Committee consisting of representative nominated by the two Government in equal numbers (hereinafter called Joint Committee) shall be constituted following the signing of this Treaty. The Joint Committee shall set up suitable teams at Farakka and Hardinage Bridge to observe and record at Farakka the daily flows below Farakka Barrage, in the Feeder Canal, and at the Navigation Lock, as well as at the Hardinage Bridge.

Article V

The joint Committee shall decide its own procedure and method of functioning.

Article VI

The Joint Committee shall submit to the two Governments all data collected by it and shall also submit a yearly report to both the Governments. Following submission of the reports the two Governments will meet at appropriate levels to decide upon such further actions as may be needed.

Article VII

The Joint Committee shall be responsible for implementing the arrangements contained in this Treaty and examining any difficulty arising out of the implementation of the above arrangements and of the operation of Farakka Barrage. Any difference of dispute arising in this regard, if not resolved by Joint Committee, shall be referred to the Indo-Bangladesh Joint River Commission. If the difference or dispute still remains unresolved, it shall be referred to the two Governments which shall meet urgently at the appropriate level to resolve it by mutual discussion.

Article VIII

The two Governments recognize the need to cooperate with each other in finding a solution to the long-term problem of augmenting the flows of the Ganga/Ganges during the dry season.

Article IX

Guided by the principles of equity, fairness and no harm to either party, both the Governments agree to conclude water sharing Treaties/Agreement with regard to other common rivers.

Article X

The sharing arrangement under this Treaty shall be reviewed by the two Governments at the five years interval or earlier, as required by either party and needed adjustments, based on principles of equity, fairness and no harm to either party made thereto, if necessary. It would be open to either party to seek the first review after two years to assess the impact and working of the sharing arrangement as contained in the Treaty.

Article XI

For the period of this Treaty, in the absence of mutual agreement on adjustments following reviews as mentioned in Article X, India shall release downstream of Farakka Barrage, water at a rate not less than 90% (ninety percent) of Bangladesh's share according to the formula referred to in Article II, until such time as mutually agreed flows are decided. This Treaty shall entered into force upon signature and shall remain in force for a period of thirty years and it shall be renewable on the basis of mutual consent.

Article XII

This Treaty shall enter into force upon signature and shall remain in force per a period of thirty years and it shall be renewable on the basis of mutual consent.

IN WITNESS WHEREOF the undersigned, being duly authorized thereto by the respective Governments, have signed this Treaty.

DONE at new Delhi 12th December,1996 in Hindi-Bangla and English languages. In the event of any conflict between the texts, the English text shall prevail.

Sd,(H.D.DEVA GOWDA)   Sd.(SHEIKH HASINA)
PRIME MINISTER   PRIME MINISTER
REPUBLIC OF INDIA   PEOPLE'S REPUBLIC OF BANGLADESH

Annexure I

Availability at Farakka	Share of India	Share of Bangladesh
70,000 cusecs or less	50%	50%
70,000 cusecs-75, 000 cusecs	Balance of flow	35,000 cusecs
75,000 cusecs or more	40,000 cusecs	Balance of flow

Subject to the condition that India and Bangladesh each shall receive guaranteed 35,000 cusecs of water in alternate three 10-day periods during the period March 11 to May 10.

Annexure II

Schedule (Sharing of waters at Farakka between January 01 and May 31 every year). If actual availability correspondents to average flows of the period 1949 to 1988, the implication of the formula in Annexure-I for the share of each side is:

Period	Average of total flow 1949–88 (cusecs)	India's share (cusecs)	Bangladesh's share (cusecs)
January			
1–10	107,516	40,000	67,516
11–20	97,673	40,000	57,673
21–31	90,154	40,000	50,154
February			
1–10	86,323	40,000	46,323
11–20	82,859	40,000	42,859
21–28	79,106	40,000	39,106
March			
1–10	74,419	39,419	35,000
11–20	68,931	33,931	35,000*
21–31	64,688	35,000*	29,688
April			
1–10	63,180	28,180	35,000*
11–20	62,663	35,000*	27,663
21–30	60,992	25,922	35,000*

Period	Average of total flow 1949–88 (cusecs)	India's share (cusecs)	Bangladesh's share (cusecs)
May			
1–10	67,351	35,000*	32,351
11–20	73,590	38,590	35,000
21–31	81,854	40,000	41,854

(* Three ten day periods during which 35,000 cusecs shall be provided)

Source: Ministry of Jal Shakti, Department of Water Resources, River Development & Ganga Rejuvenation (1996), Government of India, http://jalshakti-dowr.gov.in/sites/default/files/Indo%20Bangladesh%20Ganga%20Water%20TREATY-1996.pdf.Accessed on 19 July 2019

Appendix 3

Treaty between his Majesty's Government
of Nepal and the Government of India
concerning the integrated development
of the Mahakali Barrage including Sarada
Barrage, Tanakpur Barrage
and Pancheshwar Project

His Majesty's Government of NEPAL and the Government of INDIA (hereinafter referred to as the "Parties").

Reaffirming the determination to promote and strengthen their relations of friendship and close neighbourliness for the co-operation in the development of water resources.

Recognizing that the Mahakali River is a boundary river on major stretches between the two countries.

Realizing the desirability to enter into a treaty on the basis of equal partnership to define their obligations and corresponding rights and duties thereto in regard to the waters of the Mahakali River and its utilization.

Noting the Exchange of Letters of 1920 through which both the Parties had entered into an arrangement for the construction of the Sarada Barrage in the Mahakali River, whereby Nepal is to receive some waters from the said Barrage.

Recalling the decisions taken in the Joint Commission dated 4-5 December, 1991 and the Joint Communiqué issued during the visit of the Prime Minister of India to Nepal on 21st October, 1992 regarding the Tanakpur Barrage which India has constructed in a course of the Mahakali River with a part of the eastern afflux bund at Jimuwa and the adjoining pondage area of the said Barrage lying in the Nepali territory;

Noting that both the parties are jointly preparing a Detailed Project Report of the Pancheshwar Multipurpose Project to be implemented in the Mahakali River;

Now, therefore, the Parties hereto hereby have agreed as follows:

Article 1

1   Nepal shall have the right to a supply of 28.35 m³/s (1000 cusecs) of water from the Sarada Barrage in the wet season (i.e. from 15th May to 15th October) and 4.25 m³/s (150 cusecs) in the dry season (i.e. from 16th October to 14th May).

2   India shall maintain a flow of not less than 10 m³/s (350 cusecs) downstream of the Sarada Barrage in the Mahakali River to maintain and preserve the river eco-system.

3   In case the Sarada Barrage becomes non-functional due to any cause:

   (a) Nepal shall have the right to a supply of water as mentioned in Paragraph 1 of Article, by using the head regulator (s) mentioned in Paragraph 2 of Article 2 herein. Such a supply of water shall be in addition to the water to be supplied to Nepal pursuant to Paragraph 2 of Article 2.

   (b) India shall maintain the river flow pursuant to Paragraph 2 of this Article from the tailrace of the Tanakpur Power Station downstream of the Sarada Barrage.

Article 2

In continuation of the decisions taken in the Joint Commission dated 4-5 December, 1991 and the Joint Communiqué issued during the visit of the Prime Minister of India to Nepal on 21st October, 1992, both the Parities agree as follows:

1   For the construction of the eastern afflux bund of the Tanakpur Barrage, a Jimuwa and tying it up to the high ground in the Nepali territory at EL 250 M, Nepal gives its consent to use a piece of land of about 577 metres in length (an area of about 2.9 hectares) of the Nepali territory at the Jimuwa Village in Mahendranagar Municipal area and a certain portion of the No-Man's Land on either side of the border. The Nepali land consented to be so used and the land lying on the west of the said land (about 9 hectares) upto the Nepal-India border which forms a part of the pondage area, including the natural resources endowment lying within that area, remains under the continued sovereignty and control of Nepal and Nepal is free to exercise all attendant rights thereto.

   In lieu of the eastern afflux bund of the Tanakpur Barrage, at Jimuwa thus constructed, Nepal shall have the right to:

   (a) A supply of 28.35 m³/s (1000 cusecs) of water in the wet season (i.e. from 15th May to 15th October) and 8.50.m³/s (300 cusecs) in the dry seasons (i.e. from 16th October to 14th May) from the date of the entry into force of this Treaty. For this purpose and for the purposes of Article 1, herein India shall construct the head regulator (s) near the left under sluice of the Tanakpur Barrage and also the waterways of the required capacity upto the Nepal-India border. Such head regulator(s) and waterways shall be operated jointly.

   (b) A supply of 70 millions kilowatt-hour (unit) of energy on a continuous basis annually, free of cost, from the date of the entry into force of this Treaty. For this purpose, India shall construct a 132 KV transmission line upto the Nepal-India border from the Tanak-pur Power Station (which has, at present, an installed capacity of 120,000 kilowatt generating 448.4 millions kilowatt-hour of energy annually on 90 percent dependable year flow)

2   Following arrangements shall be made at the Tanakpur Barrage at the time of development of any storage project(s) including Pancheshwar Multipurpose Project upstream of the Tanakpur Barrage:

    (a)   Additional head regulator and the necessary waterways, as required, up to the Nepal-India border shall be constructed to supply additional water to Nepal. Such head regulator and waterways shall be operated jointly.

    (b)   Nepal shall have additional energy equal to half of the incremental energy generated from the Tanakpur Power Station, on a continuous basis from the date of augmentation of the flow of the Mahakali River and shall bear half of the additional operation cost and, if required, half of the additional capital cost at the Tanakpur Power Station for the generation of such incremental energy.

Article 3

Pancheshwar Multipurpose Project (hereinafter referred to as the "Project") is to be constructed on a stretch of the Mahakali River where, it forms the boundary between the two countries and hence both the parties agree that they have equal entitlement in the utilization of the waters of the Mahakali River without prejudice to their respective existing consumptive uses of the waters of the Mahakali River. Therefore, both the Parties agree to implement the Project in the Mahakali River in accordance with the Detailed Project Report (DPR) being jointly prepared by them. The Project shall be designed and implemented on the basis of the following principles:

1   The Project shall, as would be agreed between the parties, be designed to produce the maximum total net benefit. All benefits accruing to both the Parties, with the development of the Project in the forms of power, irrigation, flood control etc., shall be assessed.

2   The Project shall be implemented or caused to be implemented as an integrated project including power stations of equal capacity on each side of the Mahakali River. The two power stations shall be operated in an integrated manner and the total energy generated shall be shared equally between the Parties.

3   The cost of the Project shall be borne by the Parties in proportion to the benefits accruing to them. Both the Parties shall jointly endeavour to mobilize the finance required for the implementation of the Project.

4   A portion of Nepal's share of energy shall be sold to India. The quantum of such energy and its price shall be mutually agreed upon between the Parties.

Article 4

India shall supply 10m³/s (350 cusecs) of water for the irrigation of Dodhara – Chandani area of Nepali Territory. The technical and other details will be mutually worked out.

Article 5

1 Water requirements of Nepal shall be given prime consideration in the utilization of the waters of the Mahakali River.
2 Both the parties shall be entitled to draw their share of waters of the Mahakali River from the Tanakpur Barrage and/or other mutually agreed points as provided for in this Treaty and any subsequent agreement between the Parties.

Article 6

Any project, other than those mentioned herein, to be developed in the Mahakali River, where it is a boundary river, shall be designed and implemented by an agreement between the Parties on the principles established by this Treaty.

Article 7

In order to maintain the flow and level of the waters of the Mahakali River, each Party undertakes not to use or obstruct or divert the waters of the Mahakali River adversely affecting its natural flow and level except by an agreement between the Parties. Provided, however, this shall not preclude the use of the waters of the Mahakali River by the local communities living along both sides of the Mahakali River, not exceeding five (5) percent of the average Annual flow at Pancheshwar.

Article 8

This Treaty shall not preclude planning, survey, development and operation of any work on the tributaries of the Mahakali River to be carried out independently by each Party in its own territory without adversely affecting the provision of Article 7 of this Treaty.

Article 9

1 There shall be a Mahakali River Commission (hereinafter referred to as the "Commission"). The Commission shall be guided by the principles of equality, mutual benefit and no harm to either Party.
2 The Commission shall be composed of equal number of representatives from both the Parties.

3   The functions of the Commission shall, inter-alia include the following:

    a) To seek information on and, if necessary, inspect all structures included in the Treaty and make recommendations to both the Parties to take steps which shall be necessary to implement the provisions of this Treaty.

    b) To make recommendations to both the Parties for the conservation and utilization of the Mahakali River as envisaged and provided for in this Treaty.

    c) To provide expert evaluation of projects and recommendations thereto.

    d) To co-ordinate and monitor plans of actions arising out of the implementation of this Treaty; and

    e) To examine any differences arising between the Parties concerning the interpretation and application of this Treaty.

4   The expenses of the Commission shall be borne equally by both the Parties.

5   As soon as the Commission has been constituted pursuant to paragraphs 1 and 2 of this Article, it shall draft its rules of procedure which shall be submitted to both the Parties for their concurrence.

6   Both the Parties shall reserve their rights to deal directly with each other on matters which may be in the competence of the Commission.

Article 10

Both the Parties may form project specific joint entity/ies for the development, execution and operation of new projects including Pancheshwar Multipurpose Project in the Mahakali River for their mutual benefit.

Article 11

1   If the Commission fails under Article 9 of this Treaty, to recommend its opinion after examining the differences of the Parties within three (3) months of such reference to the Commission or either Party disagrees with the recommendation of the Commission, then a dispute shall be deemed to have been arisen which shall then be submitted to arbitration for decision. In so doing either Party shall give three (3) months prior notice to the other Party.

2   Arbitration shall be conducted by a tribunal composed of three arbitrators. One arbitrator shall be nominated by Nepal, one by India, with neither country to nominate its own national and the third arbitrator shall be appointed jointly, who, as a member of the tribunal, shall preside over such tribunal. In the event that the Parties are unable to agree upon the third arbitrator within ninety (90) days after receipt of a proposal, either Party may request the Secretary-General of the Permanent

Court of Arbitration at The Hague to appoint such arbitrator who shall not be a national of either country.

3   The procedures of the arbitration shall be determined by the arbitration tribunal and the decision of a majority of the arbitrators shall be the decision of the tribunal. The proceedings of the tribunal shall be conducted in English and the decision of such a tribunal shall be in writing. Both the Parties shall accept the decision as final, definitive and binding.

4   Provision for the venue of arbitration, the administrative support of the arbitration tribunal and the remuneration and expenses of its arbitrator's shall be as agreed in an exchange of notes between the Parties. Both the Parties may also agree by such exchange of notes on alternative procedures for settling differences arising under this Treaty.

Article 12

1   Following the conclusion of this Treaty, the earlier understanding reached between the Parties concerning the utilization of the water of the Mahakali River from the Sarada Barrage and the Tanakpur Barrage, which have been incorporated herein, shall be deemed to have been replaced by this Treaty.

2   This Treaty shall be subject to ratification and shall enter into force on the date of exchange of instruments of ratification. It shall remain valid for a period of seventy five (75) years from the date of its entry into force.

3   This Treaty shall be reviewed by both the Parties at ten (10) years interval or earlier as required by either Party and make amendments thereto, if necessary.

4   Agreements, as required, shall be entered into by the Parties to give effect to the provisions of this Treaty.

IN WITNESS WHEREOF the undersigned being duly authorized thereto by their respective governments have hereto signed this Treaty and affixed thereto their seals in two originals each in Hindi, Nepali and English Languages, all the texts being equally authentic. In case of doubt, the English text shall prevail.

Done at New Delhi, India, on the twelfth day of February of the year one thousand nine hundred ninety six.

Ministry of Jal Shakti, Department of Water Resources, River Development & Ganga Rejuvenation, Government of India, http://mowr.gov.in/sites/default/files/MAHAKALI_TREATY_19961.pdf. Accessed on 18 July 2019.

Appendix 4

Agreement between the Government of the Republic of India and the Royal Government of Bhutan concerning cooperation in the field of hydroelectric power

The Government of India and the Royal Government of Bhutan bearing in mind the friendly relations existing between the two countries and their people;

Recognizing that the cooperation the two countries have developed in the hydropower sector is capable of contributing greatly to the economic growth and to the greater welfare of the two countries;

Recognizing the need for energy security of their respective countries;

Being desirous of achieving the development of hydropower in the Kingdom of Bhutan in a manner that will make a lasting contribution to the economic development of both countries;

Recognizing that desirable benefits to their countries can be secured by cooperative measures for hydroelectric power generation; and

Realizing the need to accelerate construction of hydropower plants

HAVE AGREED AS FOLLOWS:

Article 1

The Government of India and the Royal Government of Bhutan agree to facilitate, encourage and promote the development and construction of hydropower projects and associated transmission systems as well as trade in electricity between the two countries, both through public and private sector participation.

Article 2

For projects to be implemented jointly by the two Governments through joint ventures or government owned agencies, a suitably empowered joint group would be set up to facilitate identification of projects, preparation of Detailed Project Reports (DPRs) and selection of agencies for speedy implementation of projects. The Government of India agrees to a minimum import of 5000 MW electricity from Bhutan by the year 2020.

Article 3

The development of projects would be governed by project implementation and power purchase agreements. The parties entering into such agreements shall mutually determine the terms and conditions of such agreements including implementation, the quantum and parameters of supply, the points of delivery and the price of supply of electrical power.

Article 4

The parties entering into such agreements shall be accorded all necessary assistance by the respective Governments, in accordance with the laws and regulations of the respective countries, for conduct of surveys including field investigations and for construction, installation, operation and maintenance of facilities required for generation, transmission, and sale of power in the territories of their respective countries.

Article 5

The parties entering into such agreements shall be granted all the incentives and concessions by the respective Governments, available under relevant laws, policies and regulations of the respective countries, for generation and transmission of power.

Article 6

The parties entering into such agreements shall fulfill all necessary requirements stipulated in relevant laws and regulations and shall also comply with necessary technical requirements of the respective countries.

Article 7

Power exchange between the two countries shall play an important role to ensure energy security. In this respect, India shall continue to supply power to Bhutan in the event of any shortfall of power supply in Bhutan through various arrangments and delivery points mutually agreed upon.

Article 8

The two countries shall cooperate in the development of renewable energy and both countries shall support each other to develop projects under the Clean Development Mechanism of the Kyoto Protocol, using India's carbon emission baseline, and any other international mechanisms that may come into force to encourage renewable energy.

Article 9

For the construction and operation of hydro projects in Bhutan by public sector and government owned agencies, India shall facilitate the availing of facilities including financing from various financial institutions in India and deployment of human resources as may be desired by Bhutan.

Article 10

Any differences regarding interpretation and application of this Agreement shall be resolved by mutual consultations between the two Governments.

Article 11

This Agreement shall come into force on the date of signing of this Agreement. It shall remain valid for a period of sixty (60) years from the date of its coming into force and its validity may be extended by mutual consent. The provisions of this Agreement shall be reviewed after an interval of ten (10) years or earlier as required by either Government and shall be amended by mutual consent. Either party desirous of terminating this Agreement shall provide a minimum of one (1) year prior notice.

Article 12

The quantum of electricity to be imported from Bhutan shall be appropriately enhanced as required during the review period.

IN WITNESS WHEREOF the undersigned, being duly authorized by their respective Governments have signed this agreement at New Delhi this Twenty Eighth Day of July, Two Thousand and Six, in two (2) original copies in Hindi, Dzongkha and English and all texts being equally authentic. In case of any divergence in interpretation, the English text shall prevail.

FOR THE GOVERNMENT   FOR THE ROYAL GOVERNMENT
OF THE REPUBLIC OF INDIA   OF BHUTAN
Sushil Kumar Shinde   Lyonpo Yeshey Zimba

Source: Available at www.internationalrivers.org/sites/default/files/attached-files/india_bhutan_hydropower_agreement_july_2006.pdf and www.internationalrivers.org/node/8375

Appendix 5

Protocol to the 2006 agreement between the Government of Republic of India and the Royal Government of Bhutan concerning cooperation in the field of hydroelectric power

The Government of India (GoI) and the Royal Government of Bhutan (RGoB) (hereinafter referred to as Parties),

PURSUANT TO the Agreement between the Government of the Republic of India and the Royal Government of Bhutan concerning cooperation shown in the development of the hydro-electric resources of Bhutan;

NOTING THE successful completion of the Chhukha Hydro-electric, the Kurichhu Hydro-electric and the Tala Hydro-electric Projects and the successful commencement of the Puantsangchhu-I Project which are all milestones in India-Bhutan cooperation in the field of hydropower; and

NOTING FURTHER the desire of the Royal Government of Bhutan to accelerate the development of additional hydropower potential in the Kingdom, and the Government of India's willingness to cooperate with and assist the Royal Government of Bhutan in attaining this objective,

Have agreed as follows:

Article I

The Government of India agrees to assist the Royal Government of Bhutan in developing a minimum of 10,000 Megawatts of hydropower and import the surplus electricity from this to India by the year 2020. The modalities of developing this hydropower would be through existing models of direct Government of India assistance and through Indian and Bhutanese Public Sector Undertakings in mutual consultation.

Article II

The Royal Government of Bhutan shall identify the priority projects to arrive at the 10,000 MW capacity in mutual consultation with the Government of India. The Government of India and the Royal Government of Bhutan shall facilitate and expedite the preparation of their Detailed Project Reports (DPRs), and their implementation.

Article III

The two Governments may designate one or more authorized agencies for trade in power across the border on a case by case basis. The designated agency(ies) shall determine protocols or specific bilateral instruments according to the needs of the Parties.

Article IV

(a) An Empowered Joint Group (EJG) for hydropower development to fast track the approval of the implementation modalities, financing mechanisms, fund flows, contingency plans and monitoring of the progress of all activities related to the preparation of DPRs and construction of the selected hydropower projects, shall be formed comprising the following:

    (i) From GoI

        i) Financial Advisor, Ministry of External Affairs, GoI
        ii) Joint Secretary (North), Ministry of External Affairs, GoI
        iii) Joint Secretary (Hydro), Ministry of Power, GoI

    (ii) RGoB shall nominate four (4) members for the EJG
    (iii) The two Governments shall nominate any other member (s) as required.

(b) The EJG shall meet in India and Bhutan once every quarter to expedite the implementation of the hydropower projects.

(c) The Minister, Ministry of Economic Affairs, RGOB shall be the Chairperson of the EJG. The EJG shall draw up its own rules of procedures.

(d) The Ambassaors of Bhutan to India and India to Bhutan shall be permanent invitees at the meetings of the EJG.

(e) As and when required, the EJG shall invite

    i) The relevant CEO(s)/MD(s) of the Special Purpose Vehicles (SPV) formed for implementation of each project;
    ii) For joint venture projects, the CMDs/Chairpersons of the equity partners of the SPV; and
    iii) Such other technical and financial experts as the EJG may deem necessary.

Article V

All other provisions of the 2006 Agreement between the Government of the Republic of India and the Royal Government of Bhutan concerning cooperation in the field of hydroelectric power shall remain unchanged.

Article VI

The two governments through mutual consultation may amend this Protocol from time to time.

IN WITNESS WHEREOF the following representatives duly authorized thereto by their respective Governments have signed this Protocol.

Signed on the Sixteenth Day in the Month of March 2009 in New Delhi in two originals in the English language.

For Government of India For   Royal Government of Bhutan
   Shri V.S. Sampath   Dasho Sonam Tshering

Source: Available at www.internationalrivers.org/sites/default/files/attached-files/india_bhutan_hydropower_protocol_march_2009.pdf and a www.internationalrivers.org/node/8375

Index

Note: Page numbers in *italic* indicate a figure and page numbers in **bold** indicate a table on the corresponding page